OUTSOURCING MANAGEMENT FUNCTIONS FOR THE ACQUISITION OF FEDERAL FACILITIES

Committee on Outsourcing Design and Construction-Related
Management Services for Federal Facilities

Board on Infrastructure and the Constructed Environment

Commission on Engineering and Technical Systems

National Research Council

NATIONAL ACADEMY PRESS
WASHINGTON, D.C.

NATIONAL ACADEMY PRESS 2101 Constitution Avenue, N.W. Washington, D.C. 20418

NOTICE: The project that is the subject of this report was approved by the Governing Board of the National Research Council, whose members are drawn from the councils of the National Academy of Sciences, the National Academy of Engineering, and the Institute of Medicine. The members of the committee responsible for the report were chosen for their special competencies and with regard for appropriate balance.

This study was supported by Contract/Grant No. S-FBOAD-94-C-0023 between the National Academy of Sciences and the Federal Facilities Council via the U.S. Department of State. Any opinions, findings, conclusions, or recommendations expressed in this publication are those of the author(s) and do not necessarily reflect the views of the organizations or agencies that provided support for the project.

Library of Congress Cataloging-in-Publication Data

International Standard Book Number 0-309-07267-0
Library of Congress Catalog Card Number 00-110532

Additional copies of this report are available for sale from National Academy Press, 2101 Constitution Avenue, N.W., Lockbox 285, Washington, D.C. 20055; (800) 624-6242 or (202) 334-3313 (in the Washington metropolitan area); also available on line at: *http://www.nap.edu*

Printed in the United States of America

THE NATIONAL ACADEMIES

National Academy of Sciences
National Academy of Engineering
Institute of Medicine
National Research Council

The **National Academy of Sciences** is a private, nonprofit, self-perpetuating society of distinguished scholars engaged in scientific and engineering research, dedicated to the furtherance of science and technology and to their use for the general welfare. Upon the authority of the charter granted to it by the Congress in 1863, the Academy has a mandate that requires it to advise the federal government on scientific and technical matters. Dr. Bruce M. Alberts is president of the National Academy of Sciences.

The **National Academy of Engineering** was established in 1964, under the charter of the National Academy of Sciences, as a parallel organization of outstanding engineers. It is autonomous in its administration and in the selection of its members, sharing with the National Academy of Sciences the responsibility for advising the federal government. The National Academy of Engineering also sponsors engineering programs aimed at meeting national needs, encourages education and research, and recognizes the superior achievements of engineers. Dr. William A. Wulf is president of the National Academy of Engineering.

The **Institute of Medicine** was established in 1970 by the National Academy of Sciences to secure the services of eminent members of appropriate professions in the examination of policy matters pertaining to the health of the public. The Institute acts under the responsibility given to the National Academy of Sciences by its congressional charter to be an adviser to the federal government and, upon its own initiative, to identify issues of medical care, research, and education. Dr. Kenneth I. Shine is president of the Institute of Medicine.

The **National Research Council** was organized by the National Academy of Sciences in 1916 to associate the broad community of science and technology with the Academy's purposes of furthering knowledge and advising the federal government. Functioning in accordance with general policies determined by the Academy, the Council has become the principal operating agency of both the National Academy of Sciences and the National Academy of Engineering in providing services to the government, the public, and the scientific and engineering communities. The Council is administered jointly by both Academies and the Institute of Medicine. Dr. Bruce M. Alberts and Dr. William A. Wulf are chairman and vice chairman, respectively, of the National Research Council.

COMMITTEE ON OUTSOURCING DESIGN AND CONSTRUCTION-RELATED MANAGEMENT SERVICES FOR FEDERAL FACILITIES

HENRY L. MICHEL, *chair*, Parsons Brinckerhoff, New York, New York
JOSEPH A. AHEARN, CH2M Hill, Greenwood Village, Colorado
A. WAYNE COLLINS, Arizona Department of Transportation, Phoenix
JOHN D. DONAHUE, Harvard University, Cambridge, Massachusetts
LLOYD A. DUSCHA, consulting engineer, Reston, Virginia
G. BRIAN ESTES, consulting engineer, Williamsburg, Virginia
MARK C. FRIEDLANDER, Schiff, Harden, and Waite, Chicago, Illinois
HENRY J. HATCH, American Society of Civil Engineers, Reston, Virginia
STEPHEN C. MITCHELL, Lester B. Knight and Associates, Inc., Chicago, Illinois
KARLA SCHIKORE, consultant, Petaluma, California
E. SARAH SLAUGHTER, MOCA, Inc., Newton, Massachusetts
LUIS M. TORMENTA, The LIRO Group, New York, New York
RICHARD L. TUCKER, University of Texas at Austin
NORBERT W. YOUNG, JR., McGraw-Hill Companies, New York, New York

Staff

RICHARD G. LITTLE, director, Board on Infrastructure and the Constructed Environment
LYNDA L. STANLEY, study director
JOHN A. WALEWSKI, project officer
LORI J. VASQUEZ, administrative associate
NICOLE E. LONGSHORE, project assistant

BOARD ON INFRASTRUCTURE AND THE CONSTRUCTED ENVIRONMENT

Acknowledgments

This report has been reviewed in draft form by individuals chosen for their diverse perspectives and technical expertise, in accordance with procedures approved by the NRC's Report Review Committee. The purpose of this independent review is to provide candid and critical comments that will assist the institution in making its published report as sound as possible and to ensure that the report meets institutional standards for objectivity, evidence, and responsiveness to the study charge. The review comments and draft manuscript remain confidential to protect the integrity of the deliberative process. We wish to thank the following individuals for their review of this report:

John Cable, University of Maryland
Frank Camm, RAND Corporation
G. Edward Gibson, University of Texas-Austin
Theodore Kennedy, BE&K, Inc.
Donald Kettl, University of Wisconsin-Madison
David Skiven, General Motors Corporation
Ralph Spillinger, Facility Management Consultant

Although the reviewers listed above have provided many constructive comments and suggestions, they were not asked to endorse the conclusions or recommendations nor did they see the final draft of the report before its release. The review of this report was overseen by RADM Donald G. Iselin, U.S. Navy (retired), appointed by the Commission on Engineering and Technical Systems, who was responsible for making certain that an independent examination of this report was carried out in accordance with institutional procedures and that all review comments were carefully considered. Responsibility for the final content of this report rests entirely with the authoring committee and the institution.

Contents

Tables and Figures

FIGURES

TABLES

Executive Summary

In this study *outsourcing* is defined as the organizational practice of contracting for services from an external entity while retaining control over assets and oversight of the services being outsourced. In the 1980s, a number of factors led to a renewed interest in outsourcing. For private sector organizations, outsourcing was identified as a strategic component of business process reengineering—an effort to streamline an organization and increase its profitability. In the public sector, growing concern about the federal budget deficit, the continuing long-term fiscal crisis of some large cities, and other factors accelerated the use of privatization[1] measures (including outsourcing for services) as a means of increasing the efficiency of government.

The literature on business management has been focused on the reengineering of business processes in the context of the financial, management, time, and staffing constraints of private enterprise. The underlying premises of business process reengineering are: (1) the essential areas of expertise, or core competencies, of an organization should be limited to a few activities that are central to its current focus and future profitability, or bottom line; and (2) because managerial time and resources are limited, they should be concentrated on the organization's core competencies. Additional functions can be retained within the organization, or in-house, to keep competitors from learning, taking over, bypassing, or eroding the organization's core business expertise. Routine or noncore elements of the business can be contracted out, or outsourced, to external entities that specialize in those services.

[1]*Privatization* has been defined as any process aimed at shifting functions and responsibilities, in whole or in part, from the government to the private sector.

Public-sector organizations, in contrast, have no bottom line comparable to the profitability of a business enterprise. The missions of governmental entities are focused on providing services related to public health, safety, and welfare; one objective is to do so cost effectively, rather than profitably. Thus, public practices are often very different from private-sector practices. They entail different risks, different operating environments, and different management systems.

Private corporations and the federal government have invested billions of dollars in facilities and infrastructure to support the services and activities necessary to fulfill their respective businesses and missions. Until the corporate downsizings of the 1980s, owners of large inventories of buildings usually maintained in-house facilities engineering organizations responsible for design, construction, operations, and project management. These engineering organizations were staffed by hundreds, sometimes thousands, of architects and engineers. In the United States during the last 20 years, almost all of these engineering organizations have been reorganized, sometimes repeatedly, as a result of business process reengineering. Some organizations are still restructuring their central engineering organizations, shifting project responsibilities to business units or operating units, and outsourcing more work to external organizations.

Studies have found that many companies are uncertain about the appropriate size and role of their in-house facilities engineering organizations. Reorganizations sometimes leave owners inadequately structured to develop and execute facility projects. In many organizations, the technical competence necessary to develop the most appropriate project to meet a business need has been lost, along with the competency to execute the project effectively. Even though many owner organizations recognize that the skills required on the owner's side to manage projects has changed dramatically, they are doing little to address this issue.

Federal agencies are experiencing changes similar to those affecting private-sector owner organizations. A survey by the Federal Facilities Council found that by 1999, in nine federal agencies, in-house facilities engineering staffs had been reduced by an average of 50 percent. The loss of expertise reflected in this statistic is compounded by the fact that procurement specialists, trained primarily in contract negotiation and review rather than in design and construction, are playing increasingly greater roles in facility acquisitions.

Outsourcing is not new to federal agencies. The government has contracted for facility planning, design, and construction services for decades. Recently, however, in response to executive and legislative initiatives to reduce the federal workforce, cut costs, improve customer service, and become more businesslike, federal agencies have begun outsourcing some management functions for facility acquisitions. The reliance on nonfederal entities to provide management functions for federal facility acquisitions has raised concerns about the level of control, management responsibility, and accountability being transferred to nonfederal service providers. Outsourcing management functions has also raised concerns about some agencies' long-term ability to plan, guide, oversee, and evaluate facility acquisitions effectively.

To address these concerns, the sponsoring agencies of the Federal Facilities Council requested that the National Research Council (NRC) develop a guide, or "road map," to help federal agencies determine which management functions for planning, design, and construction-related services may be outsourced. In carrying out this charge, the NRC committee appointed to prepare this report was asked to: (1) assess recent federal experience with the outsourcing of management functions for planning, design, and construction services; (2) develop a technical framework and methodology for implementing a successful outsourcing program; (3) identify measures to determine performance outcomes; and (4) identify the organizational core competencies necessary for effective oversight of outsourced management functions while protecting the federal interest.

DETERMINING WHICH MANAGEMENT FUNCTIONS MAY BE OUTSOURCED BY FEDERAL AGENCIES

The committee reviewed federal legislation and policies related to inherently governmental functions—a critical determinant of which activities federal agencies can and cannot outsource. An *inherently governmental function* is defined as one that is so intimately related to the public interest that it must be performed by government employees. An activity not inherently governmental is defined as *commercial*. The committee concluded that, although design and construction activities are commercial and may be outsourced, management functions cannot be clearly categorized.

In the facility acquisition process, an owner's role is to establish objectives and to make decisions on important issues. Management functions, in contrast, include the ministerial tasks necessary to accomplish the task. Based on a review of federal regulations, the committee concluded that inherently governmental functions related to facility acquisitions include making a decision (or casting a vote) pertaining to policy, prime contracts, or the commitment of government funds. None of these can be construed as ministerial functions. The distinction between activities that are inherently governmental and those that are commercial, therefore, is essentially the same as the distinction between ownership and management functions.

Using Section 7.5 of the Federal Acquisition Regulations as a basis, the committee developed a two-step threshold test to help federal agencies determine which management functions related to facility acquisitions should be performed by in-house staff and which may be considered for outsourcing to external organizations. The first step is to determine whether the function involves decision making on important issues (ownership) or ministerial or information-related services (management). In the committee's opinion, ownership functions should be performed by in-house staff and should not be outsourced.

For activities deemed to be management functions, the second step of the analysis is to consider whether outsourcing the management function might unduly compromise one or more of the agency's ownership functions. If outsourcing of a management function would unduly compromise the agency's ownership role, then it should be considered a "quasi"-inherently governmental function and should not be outsourced.

Figure ES-1 is a decision framework developed by the committee for federal agencies considering outsourcing management functions for facility acquisitions. This framework recognizes the constraints of inherently governmental functions and incorporates the committee's two-step threshold for identifying ownership functions that should be performed by in-house staff and management functions that can be considered for outsourcing. The decision framework is not intended to generate definitive recommendations about which management functions may or may not be outsourced or in what combination. The decision framework is a tool to assist decision makers in analyzing their organizational strengths and weaknesses, assessing risk in specific areas based on a project's stature and sensitivity, and, at a fundamental level, questioning whether or not a management function can best be performed by in-house staff or by an external organization.

The line between inherently governmental functions and commercial activities or between ownership and management functions, can be very fine. Distinguishing between them can be difficult and may require a case-by-case analysis of many facts and circumstances.

FEDERAL EXPERIENCE WITH THE OUTSOURCING OF MANAGEMENT FUNCTIONS

The authoring committee received briefings from several federal agencies and developed and distributed a questionnaire to sponsoring agencies of the Federal Facilities Council to solicit information on their experiences with outsourcing in general and outsourcing of management functions in particular. Seven of the 13 agencies that responded to the questionnaire had outsourced some management functions for planning, design, and construction-related activities. The primary factors cited for outsourcing management functions were lack of in-house expertise and staff shortages (54 percent of responses combined); savings on project delivery time (15 percent); and, other factors, including statutory requirements (15 percent). None of the seven agencies cited cost effectiveness or deliberate downsizing as a factor in the decision to outsource management functions. Three of the seven had outsourced management functions to other federal agencies. Their experiences varied and no trends could be determined. Agencies' experiences with outsourcing management functions to the private sector were also varied, and, again, no trends could be discerned.

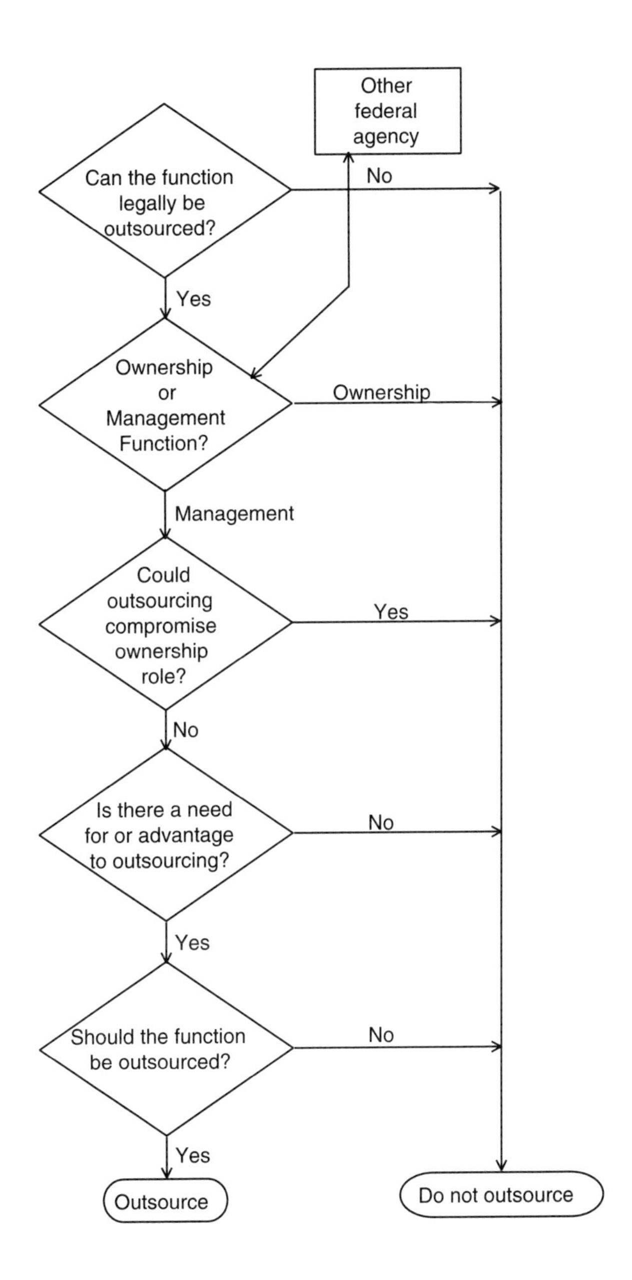

FIGURE ES-1 Decision framework for outsourcing management functions.

ORGANIZATIONAL CORE COMPETENCIES

At any one time, a federal agency may be responsible for managing several dozen to several hundred individual projects in various stages of planning, design, and construction. In some cases, agencies acquire facilities with the intent of owning and managing them directly. In other cases, agencies only require the use of facilities and may use a procuring entity to represent the government-as-owner in the acquisition process. A few agencies provide facilities for other agencies and organizations as a key component of their missions.

Core competencies constitute an organization's essential area of expertise and skill base. Unless a federal agency's mission is to provide facilities, facility acquisition and management are not core functions (i.e., facilities are not the mission but support accomplishment of the mission). However, when acquiring facilities, federal agencies assume an ownership responsibility as a steward of the public's investment. The requirements that a federal agency be accountable for upholding public policy and for committing public resources are indivisible. This combination of responsibilities requires that any federal agency that acquires facilities have the in-house capabilities to translate its mission needs directly into program definitions and project specifics and otherwise act in a publicly responsive and accountable manner. Other organizational core competencies required to direct and manage specific projects vary, depending on the agency's role as owner, user, or provider of a facility.

IMPLEMENTING A SUCCESSFUL OUTSOURCING PROGRAM

Once a decision has been made to outsource some or all management functions for facility acquisitions, the agency should clearly define the roles and responsibilities of all of the entities involved. The committee recommends that federal agencies establish and apply a responsibilities-and-deliverables matrix similar to the example shown in Figure ES-2 to help eliminate overlapping responsibilities, ensure accountability, and ensure that, as problems arise, solutions are managed effectively.

DETERMINING PERFORMANCE OUTCOMES

A key element of an organization's decision making is measuring the effectiveness of those decisions, both qualitatively and quantitatively. When management functions for facility acquisitions are outsourced, the principal measures of effectiveness of the entire effort and of individual projects should relate to cost, schedule, and safety of the projects, as well as the functionality and overall quality of the acquired facilities.

If baseline levels of service already exist or can be developed empirically, comparing the metrics and determining how well the outsourcing effort meets the

RESPONSIBILITIES-AND-DELIVERABLES MATRIX	User Management	Owner Management	Owner Project Manager	Outsourced Project Management	(A – E)	Construction Contractor	Specialty Contractors
Programming Phase							
Project request	A	P	S	S			
Deliverables/responsibilities package			A	P			
Conceptual Planning Phase							
Architect-engineer contracts			A		P		
Detailed requirements	R	A	P	S	S		
Design Phase							
Conceptual and schematic designs	R	C	A	C	P		
Permits			A	C	P		
Design development	A	A	A	C	P		
Construction documents			C	A	P		S
Procurement Phase							
List of bidders and requests for proposals			A	C	P		
Proposals (submitted)			A	C	S	P	P
Contract for construction			A	P	S	S	S
Construction Phase							
Construction permits			A	C		P	
Construction management			C	A	S	P	P
Construction work			C	A	S	P	P
Final payment (construction complete)		A	A	C	S	P	P
Start-up Phase							
Equipment installation				C	A	S	P
Move administration		P	S	S			S
Final acceptance	A	A	C	P		S	
Closeout Phase			P	S	S	S	S

FIGURE ES-2 Example of a responsibilities-and-deliverables matrix.

Note: P = primary responsibility
A = approve (signing of approval)
C = concurrence
R = reviews (no response required)
S = support (uses own resources)

basic level of expectation should be straightforward. If no baseline exists, one should be developed to ensure effective performance measurement.

Individual performance measures should be developed by the agencies that will use them and should not be prescribed by higher levels of government. Although it is entirely appropriate that operational guidance requiring the use of performance measures to be addressed be promulgated government-wide (e.g., Government Performance and Results Act) and to specify what these measures should address, the parties actually responsible for the provision of a service are in the best position to determine what constitutes good performance. Any agency that decides to outsource management functions for planning, design, and construction services should be prepared to develop and apply meaningful, measurable performance measures to determine if it is meeting its stewardship responsibilities.

FINDINGS AND RECOMMENDATIONS

The primary objective of this study is to develop a guide that federal agencies can use in the initial stages of decision making concerning the outsourcing of management functions for planning, design, and construction-related services. Agencies will have to expand and extend the guidance in this report and tailor it to their individual circumstances. By using the decision framework, by noting the findings, and by following the recommendations presented below, the committee believes federal agencies will be in a stronger position to formulate rational, business-like judgments in the public interest concerning the outsourcing of management functions for planning, design, and construction-related services.

Findings

Finding. Each federal agency involved in acquiring facilities is accountable to the U.S. government and its citizens. Each agency is responsible for managing its facilities projects and programs effectively. Responsibility for stewardship cannot be outsourced.

Finding. The outsourcing of management functions for planning, design, and construction-related services by federal agencies is a strategic decision that should be considered in the context of an agency's long-term mission.

Finding. The outsourcing of management functions for planning, design, and construction services has been practiced by some federal agencies for years. Management functions have been outsourced either to other federal agencies or the private sector. The outcomes of these efforts have varied widely, from failure to success.

Finding. At different times, an agency may fill one or more of the role(s) of owner, user, or provider of facilities.

Finding. Key factors in determining successful outcomes of outsourcing decisions include clear definitions of the scope and objectives of the services required at the beginning of the acquisition process and equally clear definitions of the roles and responsibilities of the agency. Owners and users need to provide leadership; define scope, goals, and objectives; establish performance criteria for evaluating success; allocate resources; and provide commitment and stability for achieving the goals and objectives.

Finding. Program scope, definition, and budget decisions are inherently the responsibilities of owners/users and should not be outsourced. However, assistance in discharging these responsibilities may have to be obtained by contracting for services from other federal agencies or the private sector.

Finding. The successful outsourcing of management functions by federal agencies requires competent in-house staff with a broad range of technical, financial, procurement, and management skills and a clear understanding of the agency's mission and strategic objectives.

Finding. Performance measures are necessary to assess the success of any outsourcing effort.

Finding. Because federal facilities vary widely, and because a wide range of new and evolving project delivery systems have inherently different levels of risk and management requirements, no single approach or set of organizational core competencies for the acquisition of federal facilities applies to all agencies or situations.

Finding. The organizational core competencies necessary to oversee the outsourcing of management functions for projects and/or programs need to be actively nurtured over the long term by providing opportunities for staff to obtain direct experience and training in the area of competence. The necessary skills will, in part, be determined by the role(s) the agency fills as owner, user, and/or provider of facilities.

Recommendations

Recommendation. A federal agency should analyze the relationship of outsourcing decisions to the accomplishment of its mission before outsourcing management functions for planning, design, or construction services. Outsourcing for services and functions should be integrated into an overall strategy for achieving the agency's mission, managing resources, and obtaining best value or best performance for the resources expended. Outsourcing of management functions should not be used solely as a short-term expedient to limit spending or reduce the number of in-house personnel.

Recommendation. Federal agencies should first determine their role(s) as owners, users, and/or providers of facilities and then determine the core competencies required to effectively fulfill these role(s) in overseeing the outsourcing of management functions for planning, design, and construction services.

Recommendation. Once a decision has been madc to outsource some or all management functions, a responsibilities-and-deliverables matrix should be established to help eliminate overlapping responsibilities, provide accountability, and ensure that, as problems arise, solutions are managed effectively.

Recommendation. Agencies that outsource management functions for planning, design, and construction services should regularly evaluate the effectiveness of the outsourcing effort in relation to accomplishment of the agency's mission.

Recommendation. Agencies should establish performance measures to assess accomplishments relative to the objectives established for the outsourcing effort and, at a minimum, address cost, schedule, and quality parameters.

Recommendation. Owner/user agencies should retain a sufficient level of technical and managerial competency in-house to act as informed owners and/or users when management functions for planning, design, and construction services are outsourced.

Recommendation. Provider agencies should retain a sufficient level of planning, design, and construction management activity in-house to ensure that they can act as competent providers of planning, design, and construction management services.

Recommendation. Agencies should provide training for leaders and staff responsible for technical, procurement, financial, business, and managerial functions so that they can oversee the outsourcing of management functions for planning, design, and construction effectively.

Recommendation. Interagency coordination, cooperation, collaboration, networking, and training should be increased to encourage the use of best practices and improve life-cycle cost effectiveness in the delivery of federal facilities.

1

Introduction

Outsourcing is defined in this study as the organizational practice of contracting for services from an external entity while retaining control over assets and oversight of the services being outsourced. The practice of contracting for, or outsourcing, services by industry and government is not new. The federal government, for example, has contracted with the private sector for facilities acquisition services, including planning, design, and construction services, for more than a century. In the 1980s, however, a number of factors led to a renewed interest in and emphasis on outsourcing. For private-sector organizations, outsourcing was identified as a strategic component of business-process "reengineering" designed to streamline their organizations and increase their profitability. In the public sector, growing concern about the federal government's budget deficit, the continuing, long-term fiscal crisis for some large cities, and other factors led to efforts to restrain the growth of government expenditures and accelerated the use of a wide range of privatization[1] measures, including outsourcing for services (Seidenstat, 1999).

OUTSOURCING AND BUSINESS-PROCESS REENGINEERING

The literature on business management has focused on business-process reengineering in the context of a private sector organization's financial,

[1]*Privatization* has been defined as "any process aimed at shifting functions and responsibilities, in whole or in part, from the government to the private sector" (GAO, 1997). Privatization processes may include outsourcing for services, transfer of asset ownership, managed competition/competitive contracting, franchising, public-private partnerships, vouchers, grants, subsidies or other activities (Seidenstat, 1999).

management, time, and staffing constraints. The goal of reengineering is to increase profitability by becoming more competitive and significantly improving critical areas of performance, such as quality, cost, delivery time, and customer service. The underlying premises of reengineering are: (1) the essential areas of expertise, or core competencies, of an organization should be limited to a few activities that are central to its current focus and future success, or bottom line; and (2) because managerial time and resources are limited, they should be focused on the organization's core competencies. The purpose of reengineering, then, is to streamline organizations by focusing on the core competencies required for them to compete successfully in the marketplace. Additional functions may be retained by the organization, or in-house, to keep competitors from learning, taking over, eroding, or bypassing the organization's core competencies (Pint and Baldwin, 1997).

For private sector organizations, core competencies may be skills that are: (1) difficult to duplicate; (2) create a unique value; or (3) constitute the organization's competitive advantage (i.e., what it does better than anyone else). *Core competencies* have been also been defined as a "bundle of skills and technologies that enable a company to provide a particular benefit to customers" (Hamel and Prahalad, 1994). For example, at SONY, the benefit is "pocketability, and the core competence is miniaturization," whereas at Federal Express, the benefit is on-time delivery, and the core competence is logistics management (Hamel and Prahalad, 1994).

As part of business-process reengineering, support services required by the organization that are not core competencies can be outsourced to external organizations that specialize in those services. Ideally, by outsourcing noncore functions, an organization receives the best value or best performance for the resources expended. Surveys by the Outsourcing Institute (1998) have found that private-sector organizations outsource functions for the following primary reasons:

- Improving organizational focus. By outsourcing noncore activities or operational details to an outside expert, an organization can focus its in-house resources on the development and enhancement of its core competencies.
- Gaining access to world-class capabilities. By partnering with an outside entity that has access to new technologies, tools, and techniques, an organization can gain a competitive advantage without making a substantial capital investment.
- Sharing risks. In an environment of rapidly changing markets, regulations, financial conditions, and technologies, an organization can reduce risks by sharing them with external entities.
- Reducing and controlling operating costs. By contracting with a provider that can achieve economies of scale or other cost advantages based on

specialization, an organization can reduce and control its operating expenses.

- Accelerating reengineering benefits. By outsourcing a process to an external entity that has already reengineered its business processes to world-class standards and that can guarantee the improvements and assume the risks of reengineering, an organization can realize the benefits of reengineering in less time.
- Shifting capital funds to core business areas. By reducing the need to invest in capital (building) projects or technologies by outsourcing for them, an organization can redirect capital funds to its core business activities.
- Smoothing out workloads/matching personnel to the volume of work. At times of peak business activity, an organization can contract for personnel and other resources to handle peak or unique workloads and to meet the demands of multiple projects or shifting workloads and reduce the disruptions and costs associated with hiring and then laying off "permanent" staff.

OUTSOURCING AND THE RESTRUCTURING OF GOVERNMENT SERVICES

For the federal government and its agencies, and for state and local governments, "there is no single number or 'bottom line'...comparable to the net worth of a business corporation" (OMB, 1997). In fact, there are fundamental differences between the objectives of governments and the objectives of businesses and, consequently, in the ways they operate. The primary goal of a business is to earn a profit. Government organizations, in contrast, are primarily concerned with and guided by issues of public health, safety, and welfare, such as providing national defense, conducting foreign policy, regulating goods and services, providing police and fire protection, and so on. One objective of governmental entities is to provide services cost effectively rather than for a profit. For these reasons, public practices are often very different from private sector practices; they entail different risks, different operating environments, and different management systems.

Thus, the renewed interest in outsourcing among governmental entities that began in the 1980s was generated by factors other than increasing profitability. The growing federal budget deficit, the continuing long-term fiscal crisis in large cities, grassroots efforts to restrain the growth of tax revenue and limit government spending, and efforts to make government more cost effective placed increasing pressure on elected officials at all levels of government to cut budget growth and, in some cases, downsize government operations without substantially reducing services (Seidenstat, 1999). These factors contributed to a larger national debate about which services should be directly provided by government and which should be provided by the private sector. "Once the government agenda focused on this issue of restructuring or 'rightsizing,' then privatization became a major policy option as part of a restructuring program" (Seidenstat, 1999).

As reported by Seidenstat, a 1993 report by the Council of State Governments found that, of the various privatization strategies—contracting out, grants, vouchers, volunteerism, public-private partnerships, private donations, franchises, service shedding, deregulation, and asset sales—78 percent of state agencies used contracting out as their primary strategy. Contracting out appeared to be favored because it allows: (1) government officials to retain substantial control over service production while seeking the lower costs promised by private-sector providers; (2) the public agency to have greater involvement in design, oversight, and production than other strategies; and (3) a relatively easy resumption of public production if the private provider proves to be unsatisfactory or goes out of business, particularly if the public agency retains some equipment or management expertise in the service area (Seidenstat, 1999).

A 1997 report, *Outsourcing of State Highway Facilities and Services,*[2] surveyed state transportation agencies to determine the reasons for and extent of outsourcing, among other objectives (NRC, 1997). The report found that the primary factors that influenced the decision to outsource were:

- staff constraints, including an inability to maintain or increase staff in the face of growing workloads or to meet given schedules (40 percent of total responses)
- the need for specialized expertise or equipment (24 percent)
- policy directives (22 percent)
- comparison of cost effectiveness (8 percent)
- legal requirements (4 percent)
- other factors, such as quality, the need for a neutral third party, and political or other pressures from unions or private industry (2 percent).

MAKING THE FEDERAL GOVERNMENT MORE BUSINESSLIKE

Since 1991, Congress has enacted laws and the administration has launched special initiatives to make the operations of the federal government more "businesslike" and to find an appropriate balance between functions performed by federal employees and functions performed by the private sector. This government-wide

[2]This 1997 report defined outsourcing as "contracting with either private or public-sector vendors and service suppliers to obtain services that have traditionally been, or would otherwise be, performed by staff of the state transportation agency. Subject to contractual arrangements, the responsibility to the public for the quality, reliability, and cost-effectiveness of the services may still remain with the public agency. An alternative term used to describe the same function is 'contracting out'" (NRC, 1997).

effort has been focused on improving the quality, reducing the cost, and accelerating the delivery time of services, as well as and on improving customer service; all of these objectives are shared by private sector organizations. The Government Performance and Results Act of 1993 (P.L. 103-62) mandated that federal agencies produce strategic plans and performance measures to "improve the confidence of the American people in the capability of the federal government by systematically holding federal agencies accountable for achieving program results" and to promote "a new focus on results, service quality, and customer satisfaction." The Federal Acquisition and Streamlining Act of 1994 (P.L. 103-355) and the Clingher-Cohen Act of 1996 (P.L. 104-208) were enacted to cut government red tape so that services could be provided to the public faster and more cost effectively. To meet these objectives, federal agencies were given more flexibility in finding ways to perform their missions responsibly.

During the same time period, the Base Realignment and Closure Process, the Federal Workforce Restructuring Act of 1994 (P.L. 103-226), and other legislation resulted in a substantial reduction of the federal workforce. Total federal employment declined from 2.2 million full-time equivalent positions in fiscal year (FY) 1993 to 1.9 million in FY 97 (GAO, 1999). In response to these various initiatives, federal agencies have been compelled to redefine their missions and to begin reengineering their practices and processes for conducting the business of government. Thus, although for different reasons, federal agencies, like private-sector organizations, are attempting to operate more efficiently and effectively and to obtain the best value or best performance for the resources expended.

To assess these changes, the authoring committee developed and distributed a questionnaire to sponsoring agencies of the Federal Facilities Council. Thirteen agencies responded to the questionnaire,[3] which focused on agency policies and practices related to the outsourcing of planning, design, and construction-related services (see Appendix D). All 13 agencies had outsourced planning, design, or construction-related services for 10 years or more. Seven of the 13 agencies also had experience outsourcing the management of planning, design, or construction-related services. The respondents cited the following factors for outsourcing management functions:

- lack of in-house expertise or compensation for staff shortages (54 percent of responses)
- savings on project delivery time (15 percent)
- other factors, including statutory requirements (15 percent)

---

[3]The responding agencies were the U.S. Air Force, U.S. Department of Energy, U.S. Department of State, U.S. Department of Veterans Affairs, Air National Guard, Bureau of Land Management, Bureau of Reclamation, Fish and Wildlife Service, Indian Health Service, International Broadcasting Bureau, National Institute of Standards and Technology, National Aeronautics and Space Administration, and Naval Facilities Engineering Command.

- intermittent need for services (8 percent)
- improved product quality (8 percent)

None of the seven agencies cited cost effectiveness or deliberate downsizing as a factor in outsourcing management functions.

Although this survey of federal agencies contained a smaller sample than the survey of state transportation agencies cited earlier, in both instances, staff constraints in some form were cited as the primary factor influencing the decision to outsource.

REENGINEERING OF FACILITIES ENGINEERING ORGANIZATIONS

Private corporations and the federal government have invested billions of dollars in facilities and infrastructure to support the services and activities necessary for conducting their businesses or carrying out their missions, respectively. The federal government alone owns more than 500,000 buildings and facilities worldwide valued at more than $300 billion and spends more than $20 billion per year on the design, construction, and renovation of facilities (NRC, 1998). Until the 1980s, owners of large inventories of buildings usually maintained in-house facilities engineering organizations staffed by hundreds, sometimes thousands, of architects and engineers who were responsible for designing, constructing, operating, and managing buildings, manufacturing and industrial plants, and other constructed facilities, sometimes over a wide geographical area. In the last 20 years, "nearly every owner engineering and project management organization in the United States has been reorganized, sometimes repeatedly" as a result of business process reengineering (BRT, 1997). Organizations "are restructuring their central engineering organizations, shifting project responsibilities to business units or operating facilities, and outsourcing more work to contractors" (CII, 1996). These findings are echoed in *Impact of Reengineering on Corporate Real Estate*, a 1997 benchmarking study that concluded that corporate reengineering has profound impacts on corporate real estate. "The corporate real estate department's basic mission, the services it provides, its performance and its organization and reporting are subject to intense examination. The result is often reorganization, outsourcing, performance metrics, and staff reduction" (Deloitte and Touche, 1997).

In many cases, in-house facilities engineering organizations have been reduced significantly in size and scope or eliminated altogether. One chemical firm, for example, downsized its central engineering staff from 7,000 to 1,100 positions between 1991 and 1999; a pharmaceutical firm's engineering staff was reduced from 225 to 33 engineers between 1993 and 1999 (CCIS, 1999). As part of its reengineering, an oil firm reduced its engineering staff from 1,100 to 800 positions (FFC, 1998). "Owner project skills reduced or eliminated include detailed design, project management, process engineering, construction management, technical expertise, and project controls" (CII, 1996).

The outsourcing of engineering functions is attractive to organizations because of the cyclical nature of facilities projects. Large, in-house facilities engineering groups carry substantial cost penalties when the workload is minimal, whereas large contractor firms can be hired on an as-needed basis (BRT, 1997). In addition, contractor firms might have greater expertise in new technologies, such as three-dimensional, computer-aided design software.

Sometimes, these organizational changes have unexpected results. "The engineering costs for major projects have continued to grow just as the amount of work performed in-house has declined." The reasons vary, and, "engineering costs as a percentage of total installed costs, have often not been a good indicator of project execution efficiency" (BRT, 1997). Still, "many companies are drifting because they are uncertain about the appropriate size and role of their in-house capital projects organization" (BRT, 1997). More important, reorganizations "may leave owners inadequately structured to develop and execute capital projects" because: (1) owner personnel continue to perform the same functions with fewer resources; (2) institutional knowledge is being lost through retirements; and (3) remaining personnel may not have the skills, experience, or decision-making authority necessary to perform effectively (CII, 1996). Thus, the "technical competence to assist the businesses in arriving at the most appropriate project to meet the business need has been lost along with the competence to execute the project effectively" (BRT, 1997). Recently, the Center for Construction Industry Studies found that, even though owner organizations recognize that the "skill set required to manage and work on projects from the owner's side has changed dramatically," they are "doing little to address this issue" (CCIS, 1999).

Contractor companies are also subject to organizational change as they attempt to reduce their fixed costs, increase their competitiveness, and adapt to changes in owner approaches to facilities delivery. Some contracting firms are reducing personnel despite increasing workloads, adding project skills that owner organizations are reducing or eliminating, adding skills in project phases that represent nontraditional work, and providing new services generated by advances in process, design, and construction technologies. Contractor skills that have been added include computer/software capabilities, project management, project controls, team building, project finance, project scope development, and process engineering (CII, 1996).

All of these changes have contributed to a dynamic environment. Both public and private owner organizations are trying to identify the best ways to structure their staffs and processes to acquire and renew facilities and the best ways to make owner/contractor relationships work. The Construction Industry Institute has found that "most owners do not have a process to determine owner/contractor work structure, particularly one that identifies project core competencies,

competencies to retain in-house, and those needing to be outsourced" (CII, 1996). The Business Roundtable has found that the organizations that have lost owner engineering competence eventually find the business person directly across the table from the contractors negotiating project changes, schedules, and other conditions. Because the parties do not have a common base of expertise in construction delivery, they may find it difficult to communicate (BRT, 1997). Underlying all of these issues is the critical, but often unrecognized, fact that owners and contractors have different, sometimes conflicting, goals and business objectives. Therefore, the "owner-contractor relationship needs to be structured so that each party can meet their separate goals" (CCIS, 1999).

Federal agencies, as owners of capital facilities, have been experiencing changes similar to those affecting private-sector owner organizations. A report of the Federal Construction Council noted that "due to budget cuts, agencies have had to reduce the number of project managers, design reviewers, inspectors, and field supervisors they employ" (FCC, 1987). A survey by the Federal Facilities Council found that, by 1999, in nine federal agencies, in-house facilities engineering staffs had been reduced by 20 to 65 percent, with an average of about 50 percent (FFC, 2000). An earlier study found that procurement specialists trained primarily in contract negotiation and review, rather than in design and construction, have been playing increasingly greater roles in facilities development (NRC, 1994). Thus, like private-sector facilities owners, federal agencies are faced with the challenge of identifying the essential technical and management skills that should be retained by the agency to ensure effective oversight of outsourced services.

STATEMENT OF TASK

Federal agencies have outsourced for facility planning, design, and construction services for decades. Most facility design work and almost all building construction are performed by private-sector firms. Federal agencies have traditionally relied on in-house architects and engineers to manage the design and construction of new facilities or outsourced the management to another federal agency, such as the U.S. Army Corps of Engineers (USACE), the Naval Facilities Engineering Command (NAVFAC), or the General Services Administration (GSA). Since the mid 1980s, some federal agencies have begun outsourcing some management functions for facilities acquisition, including planning, design, and construction, to the private sector. The reliance on nonfederal entities to provide management functions for federal facility acquisition has raised concerns about the level of control, responsibility, and accountability being transferred by federal agencies to nonfederal entities. Outsourcing of management functions has also raised concerns about the long-term implications for federal agencies' capabilities to plan, guide, oversee, and evaluate facility acquisitions effectively when contracting for planning, design, and construction services.

To address these concerns, the sponsoring agencies of the Federal Facilities Council[4] requested that the National Research Council (NRC) develop a guide, or road map, to help federal agencies determine which management functions for planning, design, and construction-related services may be outsourced. In response, the NRC established a committee of recognized experts in architecture, engineering, construction management, outsourcing of services, procurement and contracting, facilities management, and federal policy and procedures under the auspices of the NRC's Board on Infrastructure and the Constructed Environment (see Appendix A for biographical sketches). In carrying out its charge, the committee was asked to: (1) assess recent federal experience with the outsourcing of management functions for planning, design, and construction services; (2) develop a technical framework and methodology for implementing a successful outsourcing program; (3) identify measures to determine performance outcomes; and (4) identify the organizational core competencies necessary for effective oversight of outsourced management functions while protecting the federal interest.

The committee members, drawn from the public sector, the private sector, and academia, met three times over a 10-month period. The committee received briefings by representatives of federal agencies and private sector organizations and by invited individuals. The briefings focused on recent federal experiences with outsourcing management functions, federal policies and programs related to outsourcing and inherently governmental functions, outsourcing practices in the private sector, and related issues (see Appendix B for a list of briefings). The committee also developed and distributed a questionnaire to the sponsoring agencies of the Federal Facilities Council to gather additional information. The questionnaire focused on agency practices and experiences related to the outsourcing of planning, design, and construction services (see Appendix D for a copy of the questionnaire).

ORGANIZATION OF THIS REPORT

Chapter 2, Outsourcing Management Functions, describes how the federal government has historically acquired facilities, identifies the roles of federal agencies in acquisitions, and describes a general process and contract methods for acquiring facilities. Chapter 2 also reviews federal policies and procedures related to outsourcing and inherently governmental functions and assesses recent

---

[4]The federal agencies that contributed to the funding of this study through the Federal Facilities Council are the U.S. Air Force, Air National Guard, U.S. Army, U.S. Department of Energy, U.S. Navy, U.S. Department of State, U.S. Department of Veterans Affairs, Food and Drug Administration, General Services Administration, Indian Health Service, National Aeronautics and Space Administration, National Institutes of Health, National Institute of Standards and Technology, National Endowment for the Arts, National Science Foundation, Smithsonian Institution, and the U.S. Postal Service.

federal experiences with outsourcing management functions for planning, design, and construction services. Chapter 3, Ownership Functions and Core Competencies, focuses on owner and management functions, the characteristics and core competencies of "smart owners," "smart customers/users," and providers of facilities. Chapter 3 also addresses the development and retention of core competencies for facility acquisitions in federal agencies. Chapter 4, Decision Framework, describes a management/decision framework for considering the outsourcing of management functions for facility acquisitions and discusses the development of performance measures for evaluating the results of outsourcing.

REFERENCES

BRT (The Business Roundtable). 1997. The Business Stake in Effective Project Systems. Washington, D.C.: The Business Roundtable.

CCIS (Center for Construction Industry Studies). 1999. Owner/Contractor Organizational Changes Phase II Report. Austin, Texas: University of Texas Press.

CII (Construction Industry Institute). 1996. Owner/Contractor Work Structure: A Preview. Austin, Texas: Construction Industry Institute.

Deloitte and Touche. 1997. Impact of Reengineering on Corporate Real Estate. Chicago, Ill.: Deloitte and Touche LLP.

FCC (Federal Construction Council). 1987. Quality Control on Federal Construction Projects. Technical Report No. 84. Washington, D.C.: National Academy Press.

FFC (Federal Facilities Council). 1998. Government/Industry Forum on Capital Facilities and Core Competencies. Washington, D.C.: National Academy Press.

FFC. 2000. Adding Value to the Facility Acquisition Process: Best Practices for Reviewing Facility Designs. Washington, D.C.: National Academy Press.

GAO (General Accounting Office). 1997. Privatization: Lessons Learned by State and Local Governments. Letter Report. GGD-97-48. Washington, D.C.: Government Printing Office.

GAO. 1999. Federal Workforce: Payroll and Human Capital Changes during Downsizing. Report to Congressional Requesters. GGD-99-57. Washington, D.C.: Government Printing Office.

Hamel, G., and C.K. Prahalad. 1994. Competing for the Future. Boston, Mass.: Harvard Business School Press.

NRC (National Research Council). 1994. Responsibilities of Architects and Engineers and Their Clients in Federal Facilities Development. Board on Infrastructure and the Constructed Environment, National Research Council. Washington, D.C.: National Academy Press.

NRC. 1997. Outsourcing of State Highway Facilities and Services: A Synthesis of Highway Practice. Transportation Research Board, National Research Council. Washington, D.C.: National Academy Press.

NRC. 1998. Stewardship of Federal Facilities: A Proactive Strategy for Managing the Nation's Public Assets. Board on Infrastructure and the Constructed Environment, National Research Council. Washington, D.C.: National Academy Press.

OMB (Office of Management and Budget). 1997. Analytical Perspectives, Budget of the United States Government, Fiscal Year 1998. Washington, D.C.: Government Printing Office.

Outsourcing Institute. 1998. Survey of Current and Potential Outsourcing End-Users. Available on line at: *http://www.outsourcing.com*

Pint, E.M., and L.H. Baldwin. 1997. Strategic Sourcing: Theory and Evidence from Economics and Business Management. Prepared for the United States Air Force, Project AIR FORCE. Santa Monica, Calif.: RAND.

Seidenstat, P., ed. 1999. Contracting Out Government Services. Westport, Conn.: Praeger Publishers.

2

Outsourcing of Management Functions

One of the committee's primary tasks was to develop guidance to help federal agencies determine which management functions for planning, design, and construction services for federal facilities may be outsourced. The decision framework was developed in the context of government policies, processes, practices, and methods for acquiring facilities and related issues.

This chapter begins with a brief history of the federal organizational structure for acquiring facilities and describes a general process for facility acquisition, including process components, participants, and contract methods. Federal legislation and policies related to determining which functions are inherently governmental—a critical determinant in outsourcing decisions—are then reviewed. The chapter concludes with a brief discussion of the A-76 Process and a review of federal experiences with the outsourcing of management functions for facility acquisitions.

BRIEF HISTORY OF FEDERAL FACILITIES ACQUISITION

Federal activities related to the planning, design, and construction of facilities are as old as the nation. From the beginning of the republic through the early 1800s, government building activity increased rapidly as a result of the construction of the capital city in Washington, D.C., national expansion westward, and a growing population requiring government services. The U.S. Department of Treasury carried out most civilian building activities; USACE was primarily responsible for the construction of fortifications and other military facilities. In 1842, the Navy Bureau of Yards and Docks (now NAVFAC), was established and given authority for the design and construction of Navy yards and docks. To encourage efficient management, an Office of Construction was established in the Treasury Department in the 1850s. In 1864, the position of supervising architect was established by law to oversee the design of federal buildings (NRC, 1989).

In the post-Civil War period, federal architects and engineers continued to do much of the design work for government facilities, but the actual construction was contracted out to the private sector. The burgeoning of the government's design activities and the desire of private-sector architects to participate led to the passage of the Tarnsey Act in 1893, which permitted—but did not require—the Treasury Department to contract for private-sector architectural services selected through a competitive process. The Tarnsey Act was repealed in 1912 amid allegations of favoritism and inflated design costs.

Although the Treasury Department continued to control and execute the majority of federal building designs, by 1914 at least seven other federal agencies had in-house facilities engineering staffs for designing and managing the construction of buildings. World War I and rapid growth in the western states accelerated the trend toward the decentralization of design and construction activities. The Public Building Acts of 1926 and 1930 again authorized the Treasury Department to acquire design services from the private sector. The Great Depression accelerated demands by private-sector architects to be allowed to compete for government design work (NRC, 1989).

In 1939, the Public Buildings Administration was created within the Federal Works Agency, and it took on many of the responsibilities of the Treasury Department's Office of Construction. The title of supervising architect disappeared, and, for the first time, some design supervision was delegated to regional offices, weakening the government's centralized control over design. The National Security Act of 1947 created the Air Force and required that a majority of military construction projects be executed by the Army or the Navy. Today, except for activities reserved for readiness training, virtually all planning, design, and construction for the Air Force are managed by USACE and NAVFAC.

GSA, which was created in 1949, subsumed the Public Buildings Administration. From the beginning, the GSA has contracted out for the majority of its design and construction-related services. At the time the GSA was established, the U.S. Department of Defense, the Veterans Administration (now the U.S. Department of Veterans Affairs), the U.S. Department of State, and the National Park Service all maintained in-house capabilities in building operations to meet their own needs (NRC, 1989). See Table 2-1 for important dates in the history of federal facilities acquisition.

The decentralization of federal building activity has continued in the last 50 years. As of 1999, at least 25 separate federal entities were involved in acquiring facilities.[1] Individual federal agencies program and budget for facility acquisition

---

[1]The entities involved in acquiring planning, design, and construction-related services include, but are not limited to, the U.S. Air Force, Air National Guard, Army Corps of Engineers, U.S. Department

TABLE 2-1 Important Dates in Federal Facilities Acquisition

1800s	Federal facilities design and construction carried out by Treasury Department (civilian) and U.S. Army Corps of Engineers (military).
1842	Navy Bureau of Yards and Docks (now NAVFAC) established with authority for design and construction.
1864	Position of supervising architect established in Treasury Department.
1893	Tarnsey Act permits Treasury Department to contract for private sector architectural services.
1912	Tarnsey Act repealed.
1924, 1930	Public Building Acts authorize Treasury Department to acquire design services from the private sector.
1939	Public Buildings Administration established, absorbing responsibilities of the Treasury Department. Title of supervising architect abolished.
1947	National Security Act establishes U.S. Air Force, assigns design and construction activities to USACE and NAVFAC.
1949	General Services Administration established, assumes responsibilities of Public Buildings Administration.
1949–1999	Decentralization of design and construction continues. At least 25 separate agencies involved.

to support their mission requirements, such as military readiness, statutory compliance, and the delivery of government services. The types of facilities include hangars, warehouses, docks, military installations, office administrative space to deliver government services, courthouses, prisons, foreign embassy compounds, nuclear plants, dams, park facilities, museums, monuments, archives, laboratories, and research centers, among others. Thus, in addition to being the nation's largest owner of buildings and facilities, the federal government is also responsible for the stewardship of the most diverse facilities portfolio in the United States.

---

of Energy, Naval Facilities Engineering Command, U.S. Department of Justice, U.S. Department of Veterans Affairs, General Services Administration, Indian Health Service, National Aeronautics and Space Administration, National Institutes of Health, National Institute of Standards and Technology, Smithsonian Institution, U.S. Postal Service, National Park Service, Bureau of Indian Affairs, Internal Revenue Service, Public Health Service, U.S. Department of Agriculture, U.S. Department of Transportation, Environmental Protection Agency, Bureau of Prisons, Administrative Office of the U.S. Courts, the U.S. Coast Guard, the Architect of the Capitol, and the U.S. Department of State.

ROLES OF FEDERAL AGENCIES IN FACILITIES ACQUISITION

At any one time, a federal agency may be responsible for managing several dozen to several hundred individual projects in various stages of planning, design, and construction. As a result of relatively recent legislation and changes to the Federal Acquisition Regulations (FAR), federal agencies may acquire facilities using a variety of contracting methods: design-bid-build; design-build; construction management; program management; or variations of these. The use of performance-based specifications in contracts is also increasing.

Agency budgets for facilities acquisition vary widely. The Indian Health Service, a relatively small agency, had more than $265 million worth of building activity in planning, design, and construction as of 1999. In contrast, the U.S. Navy has an annual design and construction budget of about $2.5 billion (FFC, 2000). As missions, priorities, and situations change, the scope and budget of agency facility-acquisition programs can fluctuate greatly. For example, a recent program to upgrade federal courthouses around the country has added billions of dollars to GSA's construction budget. The U.S. Department of State is facing a similar situation. Following the 1998 bombings of U.S. embassies in Africa, legislation requiring rapid, extensive upgrades of embassy security features worldwide was enacted, which could require several billion dollars to execute (FFC, 2000).

When acquiring facilities, an agency may be acting in one or more distinct roles—as an owner, a user, or a provider of facilities. In some cases, agencies acquire facilities with the intent of owning and managing them directly. In other cases, agencies only require the use of facilities and may use a procuring entity to represent the government-as-owner in the acquisition process. Procuring entities include separate executive departments, such as GSA, or private-sector firms. A government procuring agency may also be an office or division in the same agency, such as the Office of Foreign Building Operations, which acquires facilities for use by the State Department's diplomatic staff (NRC, 1994). A few agencies, primarily GSA, USACE, and NAVFAC, provide facilities for other agencies and organizations as a key component of their missions.

GENERAL FACILITY ACQUISITION PROCESS

The federal government has not established a single, standardized process for facility acquisition, although it has established general guidance through legislation and regulations. In practice, the complex and diverse nature of federal projects, the variety of contracting methods, and the decentralization of facilities acquisitions preclude an exact, systematic, or single process for programming, budgeting, planning, designing, or constructing a facility. Within the guidance provided, agencies have developed policies, practices, criteria, and/or guidelines for facility acquisitions that reflect their missions, cultures, and resources. Thus, although agencies follow similar procedures and decision-making processes, the number, name, substance, and sequence of acquisition phases may vary. With these caveats, a general process for acquiring facilities is shown in Figure 2-1 and described below as a context for the committee's findings and recommendations.

Requirements Assessment Phase

The federal budgeting process requires that agencies conform to a procedure of setting requirements and prioritization before agency budget requests are submitted to Congress. The requirements assessment phase (also called project requirements, project assessment, or needs assessment) begins when someone (e.g., facilities program manager, senior executive, or elected official) identifies the need for a facility. In response, the agency initiates a process to gather information and validate the need for the facility relative to its mission. As part of this assessment, an agency may review its entire facilities inventory and determine whether existing buildings and infrastructure can adequately support mission and program requirements or if facilities will have to be acquired, upgraded, or replaced.

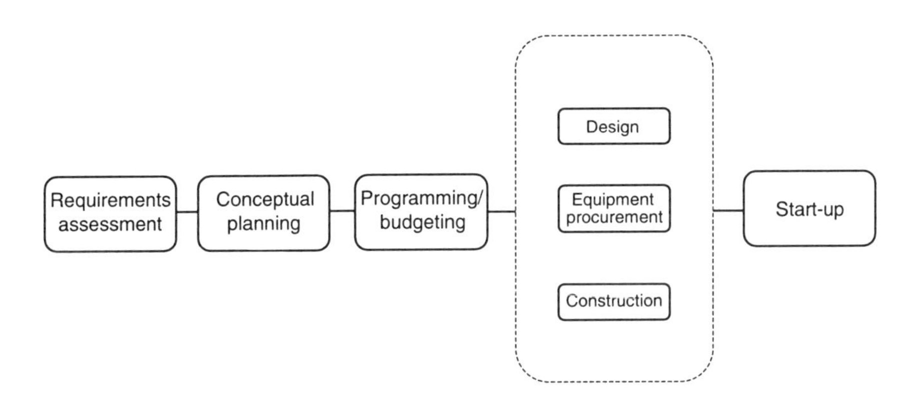

FIGURE 2-1 General facility acquisition process.

Note: The contracting method determines whether the design, equipment procurement, and construction phases occur in sequence or concurrently. The contracting method can also affect who is involved at each phase (architect, engineer, construction contractor, etc.). For example, if the design-bid-build contract method is used, the phases generally occur in sequence, with an architect-engineer entity involved in the design phase and a construction entity involved in the construction phase. If a design-build contract method is used, the same contractor is responsible for the design and construction phases; thus, some phases or activities occur concurrently.

The requirements assessment phase typically identifies space requirements by use and number of personnel. At this point, agencies that have the authority and resources to do so proceed to the conceptual planning stage. Other agencies must first prepare a request for initial congressional approval to acquire facilities and the appropriation of funds for conceptual planning and design services. The request is typically structured to meet the requirements of the particular congressional committee responsible for agency appropriations. Therefore, the exact documentation varies, but it usually includes materials justifying the facilities in relation to mission requirements and the locational, physical, and functional requirements upon which preliminary cost estimates are based.

In this phase, some agencies contract with external organizations to conduct preliminary planning and design studies that are used as the basis for formal documents submitted for congressional action (NRC, 1994). Agencies may also contract for other consultation services, such as separate opinions on long-range planning, validation of agency projections, or strategic facility planning. The decision to seek authorization and funding to acquire a particular facility to meet mission requirements, however, is the responsibility of the government agency.

Conceptual Planning Phase

In the conceptual planning phase (also called project preplanning, master planning, advance planning, front-end loading, and concept development), alternative designs are developed and considered. Functional requirements, such as floor areas for particular activities and for required or desired adjacencies and connections among activities, are developed (NRC, 1994). Various feasibility studies are conducted to define the scope or statement of work based on the agency's expectations of facility performance, quality, cost, and schedule. Several alternative design solutions may be considered during this phase, leading up to the selection of a single preferred approach that will be the basis of the scope of work used in making future decisions and in procuring design and construction services.

Studies by academicians, the NRC, the Construction Industry Institute, The Business Roundtable, and the Project Management Institute have all highlighted the importance of conceptual or advance planning to the entire facility acquisition process. Predesign phases, during which the size, function, general character, location, and budget for a facility are established, are critical. Errors at this stage are usually embodied in the completed facility in forms such as inappropriate space allocations and inadequate equipment capacity (NRC, 1989, 1999). The Business Roundtable has stated:

> The supply chain of a capital project starts with the identification of a customer need that might be translatable into a business opportunity. The front-end loading process is made up of the critical planning phases of the project. It is called front-end loading because the effective commitment of time and resources at this point dictate the future success of the project (BRT, 1997).

The importance of the conceptual planning phase is illustrated in Figure 2-2. The cost-influence curve indicates that the ability to influence the ultimate cost of a project is greatest during the conceptual planning phase and decreases rapidly as the project matures. Conversely, a project cash-flow curve shows that conceptual planning and design costs are relatively minor and that costs escalate significantly as the project evolves through the equipment procurement and construction phases.

The project scope and statement of work for a federal facility may be developed by an agency's in-house staff or with the assistance of external entities. Outside assistance might include the development of alternative design concepts or cost estimates. The responsibility for the elements included in the scope of work, however, ultimately rests with the agency.

Programming / Budgeting Phase

Once senior agency officials have determined that a project is critical to the agency's mission and, therefore, warrants acquisition, the agency prepares a request for congressional approval to acquire the facility and / or for the appropriation of funds for design and / or construction services.

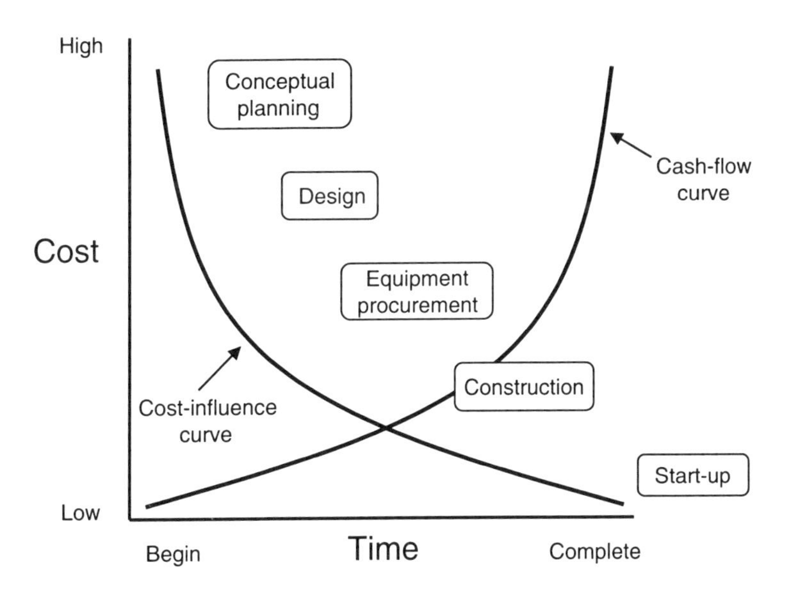

FIGURE 2-2 Cost-influence and cash-flow curves. Source: FFC, 2000.

Design Phase

Once the scope of work has been established and funds have been appropriated, the design phase begins. The preferred design approach matures into drawings, specifications, construction methods, schedules, and other documents from which equipment procurement and construction bids can be solicited.

Unless agencies have in-house staff available, design activities are typically contracted out to organizations that have the appropriate expertise, either other federal agencies that provide facilities or private-sector architect-engineer firms. Agencies may retain management and oversight responsibility by incorporating a review process to ensure that the designer accomplishes the tasks contracted for, conforms to the budget, and so forth. Agencies may also contract for related services, such as reviewing for code compliance and structural integrity, risk assessment, and compliance to design and engineering standards.

An NRC report in 1990 found that key factors in the design-related increases in construction costs that exceed budgets for federal facilities are poor planning and the failure to think carefully about foreseeable construction problems. The report also found that the early stages of the design process are critical for ensuring design to budget because, at that point, the design is still flexible, and factors that determine cost are not yet fixed (NRC, 1990).

Equipment Procurement Phase

Complex projects may include an equipment procurement phase to expedite the purchase, manufacture, and delivery of long-lead-time equipment, such as unique process machinery, large electrical and mechanical equipment, and sophisticated architectural components. Equipment procurement may proceed in parallel with construction-phase activities to ensure that long-lead-time equipment is furnished to the construction contractor at the proper time to avoid construction delays attributable to late deliveries.

Construction Phase

The construction phase of the acquisition process is by far the most costly. Approximately 70 to 90 percent of total project funding is spent during the construction phase.[2] Facility planning and design, in contrast, typically account for only 10 to 30 percent of acquisition cost, including site exploration, regulatory reviews, and other activities required prior to occupancy (NRC, 1994). A significant challenge during the construction phase is managing changes resulting from

[2]In the total life cycle of the facility (i.e., planning, design, construction, operation, maintenance, repair, renewal, disposal), design and construction costs account for 5 to 10 percent of the total costs of ownership; operation and maintenance costs account for 65 to 80 percent of total costs (NRC, 1998).

changes in the scope of work by the owner, errors and omissions in the construction documents, and changed or unknown site conditions. The construction phase is considered complete when the agency accepts occupancy of the facility, although final closeout of the construction contract may continue for months (or even years) until all discrepancies have been identified and resolved.

The majority of tasks associated with federal construction are outsourced to the private sector. Architect-engineer and construction-management firms may even be retained to advise an agency about the responsiveness of bidders to bid requirements or to assess contractors to determine if they are capable of performing the tasks that would be assigned.

Federal agencies typically use in-house personnel for construction activities only when these functions are central to the agency's mission or existence. The military services, for example, typically retain control of facility construction that affects mission readiness, although even they sometimes use private contractors in support of military deployments. Design and construction quality assurance[3] are traditionally performed by government employees, although they have also been performed by contractors.

Start-Up Phase

The start-up phase, sometimes called commissioning, begins with the initial period of occupancy of the facility by its users. A "shakedown" of building equipment occurs during this phase, during which building components are tested individually and then in conjunction with other components in the system to measure and compare their performance against the original design criteria. Facility operation and maintenance plans are implemented, tested, and refined. Minor repairs and alterations are made, and users may learn about the facility (NRC, 1994).

Start-up phase activities are often informal, but some agencies work with the facility user and building contractor during a more formal four to six month commissioning process. Specialist consultants or the architect-engineer designer may also be hired to participate in the commissioning process (NRC, 1994).

CONTRACT METHODS

Since 1994, a number of laws have been enacted to allow agencies more choices in selecting the contract method for acquiring facilities. Before the early 1990s, the design-bid-build method, described below, was used almost exclusively. Although there are variations, current federal practice recognizes four basic contract types that apply to several facility acquisition systems:

---

[3]*Quality assurance* is defined as the process of evaluating overall project performance on a regular basis to provide confidence that the project will satisfy the relevant quality standards (PMI, 1996).

- design-bid-build
- construction management
- design-build
- program management

The level of involvement and oversight by the owner organization varies depending on the contract method. Consequently, the relationship with the contractor(s) also varies.

The design-bid-build approach assumes that the owner organization contracts individually for the design, engineering, and construction services required to acquire a facility. The owner organization manages individual contracts with all design, engineering, and construction service providers, implying that the owner must also manage all interfaces between service providers. Under this approach owners commonly enlist outside consultants for various functions of the acquisition process. Interface management is critical for assessing accountability for problems incurred during the project's evolution, which can be difficult because of the variety and separation of individual contracts. This contract method requires that the owner organization maintain a relatively large and experienced in-house design, engineering, and management staff (FFC, 2000).

For the construction management approach, the owner contracts with an external entity to manage the construction of a project. The construction manager (CM) may function either as an "agency" CM or as an "at-risk" CM:

- Agency CM. The owner holds all individual construction contracts, and the CM functions as the construction contract administrator, acting on behalf of the owner and rendering an account of activities. Actual construction work is performed by others under direct contract to the owner. The CM is typically not responsible for construction means and methods and does not guarantee construction cost, time, or quality.
- At-risk CM. The actual construction work is performed by trade contractors under contract to the CM, who then becomes responsible to the owner for construction means and methods and delivery of the completed facility within the owner's scope of work for cost, time, and quality.

Under the construction management approach, the owner typically retains responsibility for managing all preconstruction architecture-engineering services and, therefore, must address all interface issues between service providers (FFC, 2000).

Under the design-build approach, an owner organization prepares a project scope definition and then engages a single entity to provide all services necessary to complete the design and construct the facility. Generally, the scope definition package represents a design that is between 15 and 35 percent complete, although variations of the design-build approach may begin much earlier, often with a performance specification, or much later, with perhaps a 65-percent design package.

Project success under the design-build approach is primarily dependent on the owner organization's ability to produce a comprehensive, well defined, unambiguous scope of work upon which all subsequent design-build activity will be based. Once the design-build contract has been awarded, changes to owner requirements generally incur heavy penalties in project cost and schedule (FFC, 2000).

For the program management approach, the owner organization contracts with a program manager (PM) to exercise oversight of the entire facility acquisition process, from planning through design, construction, outfitting, and start-up. Similar to the CM, the PM can serve in either an agency-PM or at-risk-PM capacity:

- Agency PM. The owner holds all individual contracts, and the PM functions as the contracts administrator, acting on behalf of the owner and rendering an account of activities. All project work is performed by others under direct contract to the owner. The PM is typically not responsible for project means and methods and does not guarantee cost, time, or quality.
- At-risk PM. All project work is performed by service and trade contractors under contract to the PM, who then becomes responsible to the owner for project means and methods, as well as delivery of the completed facility within the owner's objectives of cost, time, and quality.

Because the PM is responsible for managing the interfaces between all phases of facility acquisition and all parties involved, the owner organization's participation in the facility acquisition process is minimal (FFC, 2000).

INHERENTLY GOVERNMENTAL FUNCTIONS

Although the federal government has contracted with the private sector for various design and construction activities for more than a century, a larger philosophical debate over which functions should be provided by the government and performed by government employees—so-called *inherently governmental functions*—and which functions should be provided by the private sector—so-called *commercial activities* can be traced back to discussions among the framers of the Constitution that appear in the Federalist Papers (GAO, 1992). The debate continues today.

In the last 50 years, efforts have been made to develop policies and guidelines for federal agencies to determine which functions should be performed only by government employees and which ones can be performed either by government employees or private-sector contractors. Budget Bulletin No. 55-4, issued by the Eisenhower administration in January 1955, stated "it is the general policy of the Federal Government that it will not start or carry on any commercial activity to provide a service or product for its own use if such product or service can be procured from private enterprise through ordinary business channels" (Childs,

1998). Bulletins No. 57-7 and 60-2, issued in 1957 and 1962, respectively, reiterated this policy and added guidance for evaluating commercial activities, making cost comparisons, and initiating new in-house activities (Childs, 1998).

In 1966, the Office of Management and Budget (OMB) issued Circular A-76, *Performance of Commercial Activities*, which stated that "The guidelines of this Circular are in furtherance of the Government's general policy of relying on the private enterprise system to supply its needs." The circular was revised and reissued in 1979, 1983, 1996, and 1999. (In 1979, OMB issued a supplemental handbook to Circular A-76 that included detailed procedures for competitively determining whether commercial activities should be performed by in-house employees, by another federal agency through an interservice support agreement, or by the private sector. The experience of federal agencies with the so-called "A-76 process" is discussed later in this chapter.)

The 1983 version of Circular A-76 provided some guidance regarding governmental versus commercial functions and who should perform them. A governmental function was defined as:

> ...a function which is so intimately related to the public interest as to mandate performance by Government employees. These functions include those activities that require either the exercise of discretion in applying Government authority or the use of value judgment in making decisions for the government. Governmental functions normally fall into two categories: (1) the act of governing, i.e., the discretionary exercise of Government authority...(2) monetary transactions and entitlements (EOP, 1983).

An activity not considered governmental in nature was deemed a commercial activity and defined as:

> ...[a function] which is operated by a Federal executive agency and which provides a product or service which could be obtained from a commercial source....A commercial activity also may be part of an organization or a type of work that is separable from other functions or activities and is suitable for performance by contract (EOP, 1983).

To supplement OMB's guidance, some agencies, including the U.S. Department of Energy (DOE) and the Environmental Protection Agency, developed more specific guidelines for use by their employees (GAO, 1992). However, a 1990 report of the President's Council on Management Improvement focusing on Circular A-76 concluded:

> The identification of what are inherently governmental functions...and identification of commercial activities that can be contracted out is frequently contentious and difficult to accomplish. The identification process is normally unique to each organization's programs and circumstances...the definition of activities inherently governmental in nature is unclear, and there is little consensus as to which activities are governmental in nature and which are not (GAO, 1992).

The General Accounting Office (GAO), in a review of 108 randomly selected contracts issued by several federal agencies between 1989 and 1991, found that:

> The problem of defining governmental functions becomes particularly complex because consultants and management support contractors administer a broad range of activities for government agencies. Such activities involve a variety of work, such as preparing studies and analyses that are to assist agencies in making policy decisions, researching technical issues that may be beyond the expertise of available agency technical staff, developing agency reports, preparing testimony, and conducting administrative hearings (GAO, 1992).

In September 1992, the Office of Federal Procurement Policy (OFPP) issued Policy Letter 92-1 to the heads of executive agencies and departments. Policy Letter 92-1 established the policy of the executive branch on service contracting and inherently governmental functions. Its purpose was "to assist Executive Branch officers and employees in avoiding an unacceptable transfer of official responsibility to Government contractors," although the letter did "not purport to specify which functions are, as a legal matter, inherently governmental, or to define the factors used in making such a legal determination."

Policy Letter 92-1 incorporated OMB's definition of an inherently governmental function and added that:

> an inherently governmental function involves, among other things, the interpretation and execution of the laws of the United States so as to
>
> (a) bind the United States to take or not to take some action by contract, policy, regulation, authorization, order or otherwise;
>
> (b) determine, protect and advance its economic, political, territorial, property, or other interests by military or diplomatic action, civil or criminal judicial proceedings, contract management, or otherwise;
>
> (c) significantly affect the life, liberty or property of private persons;
>
> (d) commission, appoint, direct or control officers or employees of the United States; or
>
> (e) exert ultimate control over the acquisition, use, or disposition of the property, real or personal, tangible or intangible of the United States, including the collection, control, or disbursement of appropriated and other Federal funds. (OFPP Policy Letter 92-1 is reprinted in Appendix C).

In October 1998, the Federal Activities Inventory Reform (FAIR) Act of 1998 was signed (reprinted in Appendix C.) This act codifies the definition of an inherently governmental function found in OFPP Policy Letter 92-1. The FAIR Act also states that the following functions are not inherently governmental:

> (i.) gathering information for or providing advice, opinions, recommendations, or ideas to Federal Government officials; or
>
> (ii.) any function that is primarily ministerial and internal in nature (such as building security, mail operations, operation of cafeterias, housekeeping, facilities operations and maintenance, warehouse operations, motor vehicle fleet management operations, or other routine electrical or mechanical services).

Subpart 7.5, Inherently Governmental Functions of Part 7, Acquisition Planning, of the FAR implements the policies of OFPP Policy Letter 92-1. Subpart 7.5 became effective on March 26, 1996, and was reissued on December 27,

1999. The purpose of this subpart is to "prescribe policies and procedures to ensure that inherently governmental functions are not performed by contractors." Chapter 3 of this report includes a more detailed discussion of relevant elements of FAR Section 7.5; although there is no debate about design and construction functions being commercial activities, the distinction related to the management of such functions is not as clear.

THE A-76 PROCESS

As noted above, in 1979, OMB issued a supplemental handbook to Circular A-76. The handbook, amended in 1983, 1996, and 1999, provides guidance and procedures for federal agencies to determine if recurring commercial activities should be operated under contract with commercial sources, in-house using government facilities and personnel, or through interservice support agreements. The introduction to the handbook states:

> Circular A-76 is not designed to simply contract out. Rather, it is designed to: (1) balance the interests of the parties to a make or buy cost comparison, (2) provide a level playing field between public and private offerors to a competition, and (3) encourage competition and choice in the management and performance of commercial activities. It is designed to empower Federal managers to make sound and justifiable business decisions (OMB, 1999).

Under the A-76 process and related legislation, agencies are required to evaluate their activities to determine whether they are inherently governmental functions or commercial activities and to complete an inventory of commercial activities. New and expanded activities may be directly outsourced without using the A-76 process as can national defense activities, direct patient care, and other exempted activities. In certain circumstances, including those where an agency manager may want to change the method of performance and that may involve more than 10 federal staff positions, agencies must conduct cost comparisons to determine the most efficient means of carrying out the commercial activities. This involves a three-step process to determine who will perform recurring commercial activities. The first step is to develop a performance work statement defining the technical, functional, and performance characteristics of the work to be performed. The second step is to conduct a management study to determine organizational structure, staffing, and operating procedures for the most efficient organization (MEO) for effective in-house performance of the commercial activity. The third step is to accept formal bids and conduct a cost comparison between the private sector and the government's MEO to decide if an activity will be performed by government employees or the private sector (GAO, 1998).

In 1998, GAO reported that there had been "minimal A-76 activity among many agencies since the late 1980s" (GAO, 1998). Reasons cited for the lack of activity included the time and expense of conducting A-76 cost comparisons, changing management priorities, the lack of staff with the necessary technical skills to conduct the comparisons, the lack of offers from the private sector in response to solicitations, the government's lack of complete cost data (particularly for indirect costs), and limited leadership by OMB in ensuring implementation. Federal agencies established since the 1950s, including the National Aeronautics and Space Administration (NASA), DOE, the Environmental Protection Agency, and the Health Care Finance Administration, "have relied from the start on contracting out much of their work rather than performing it directly" (GAO, 1998). The GAO also noted that some agencies had done cost comparisons for providing services, but the comparisons did not involve any federal positions and, therefore, did not require A-76 analyses. Nevertheless, GAO concluded:

> Agencies' experiences with A-76 suggest that competition is a key to realizing savings, whether functions are eventually performed by private sector sources or remain in-house...there appears to be a clear consensus....that savings are possible when agencies undertake a disciplined approach, such as that called for under A-76, to review their operations and implement the changes to become more efficient themselves or contract with the private sector for services (GAO, 1998).

OUTSOURCING OF MANAGEMENT FUNCTIONS FOR FEDERAL FACILITY ACQUISITIONS

One element of the committee's statement of task was to assess recent federal experiences with the outsourcing of management functions for planning, design, and construction services. The committee received briefings on this subject from the U.S. Department of State, DOE, the U.S. Department of Veterans Affairs, and the U.S. Air Force. The committee also developed a questionnaire on this subject that was distributed to the sponsoring agencies of the Federal Facilities Council. Thirteen agencies responded. Of the 13, seven reported that they had outsourced some management functions for facility acquisition services at some time since 1980. Those agencies included the U.S. Department of State, the International Broadcasting Bureau (IBB), NASA, DOE, the National Institute of Standards and Technology (NIST), the U.S. Air Force, and the Air National Guard (ANG).[4] NAVFAC had provided management functions under contract to other federal agencies.

U.S. Department of State

The Office of Foreign Buildings Operations (FBO) of the U.S. Department of State is responsible for acquiring embassies, housing, and other facilities for

[4]The responding agencies that reported they had not outsourced management functions at the time the questionnaire was distributed were the Indian Health Service, NAVFAC, U.S. Department of Veterans Affairs, the Bureau of Reclamation, Fish and Wildlife Service, and the Bureau of Land Management.

approximately 260 diplomatic missions worldwide. FBO exercises overall responsibility for project management during the site acquisition, design, construction, and commissioning phases of acquisition, including security support. FBO develops and implements project execution plans, including project status reporting requirements, and manages the project planning system to ensure that project schedules, costs, and resources are effectively reviewed.

In the 1980s, FBO was shifting its contracting method to program management. At the same time, legislation required that a major portion of existing overseas facilities be replaced to comply with new security standards. Through a competitive bidding process, FBO outsourced for the expertise to develop a program management project delivery system and for staff services until FBO could hire permanent in-house staff. The value of the contract was approximately $76 million, and the value of the work managed was approximately $850 million. Fifty-seven task orders were issued for work on project and technical studies. Each project was subject to the development of a delivery order that specified hours, resources, and costs; tasks were identified in a series of project management manuals establishing the quality and products expected. FBO used a series of reports and management techniques to control and oversee the work of the contractor. FBO did not complete an A-76 study or establish formal performance standards or other methods of quantifying the outcomes of the outsourcing.

FBO reported that the key results of this outsourcing experience were compensation for a lack of in-house expertise and a staff shortage. In addition, "other benefits were produced that focused on project delivery times and quality of product deliverables." By the completion of the contract, FBO had hired and trained sufficient in-house staff and transferred the necessary technology to resume direct management of projects by federal employees.

The FBO response to the questionnaire also noted that outsourcing management functions to another federal agency was less successful than outsourcing to a private-sector firm. "Using another government agency with established procedures and internal administrative processes at variance with FBO created problems such as variation in standards and quality of acceptable product delivery established in each agency, and the variation in each agency's criteria for contract completion."

International Broadcasting Bureau

The Office of Engineering and Technical Operations of the IBB was responsible for the construction of radio relay stations for the U.S. Information Agency and Voice of America when the committee's questionnaire was distributed (the IBB has since been reorganized). At the time, IBB managed all planning, design, and construction project/program implementation using a combination of project managers in the Washington, D.C., headquarters office and construction and relay station managers overseas. The types of facilities acquired required highly specialized skills related to high-power, high-frequency transmitters and antennas and satellite program delivery.

In the mid-1980s, the IBB undertook an unprecedented $1 billion modernization program involving the construction of several new relay stations. A significant portion of this construction was attributable to international political changes following the end of the Cold War and required that IBB expand and update its facilities. Because IBB did not have enough in-house staff, skills, or expertise to manage the modernization program, it outsourced management functions to USACE and independent contractors. A support agreement defining the scope of work and outlining the responsibilities of each entity was written for each project. An IBB project manager was assigned to each project to oversee the work of USACE or contractor personnel. An IBB construction manager was assigned to each location where work was under way to act as a liaison between IBB and USACE and contractor personnel. Project changes were monitored, and most were submitted to the IBB for review and approval before implementation. Guidelines were issued outlining the requirements for review and/or approval of each level of project change. IBB did not complete an A-76 study or establish formal performance standards or other methods of quantifying the outcomes of the outsourcing.

IBB reported that the key results of this outsourcing experience were improved quality of the product and the growth and training of in-house personnel. In-house staff acquired the skills and expertise to manage subsequent specialized projects, thus eliminating the need for outsourcing—and reducing overall costs for these projects. IBB staff were also able to resolve day-to-day problems in operating and maintaining the relay stations, problems that could not be easily resolved through outsourcing because only a few outside organizations specialize in the installation and maintenance of high-power, high-frequency transmitters and antennas and because of the remote locations of many of these facilities.

National Aeronautics and Space Administration

NASA maintains a Facilities Engineering Division at NASA headquarters whose responsibility is to develop the Construction of Facilities Budget, obtain authority and resources, and oversee and direct NASA's 10 centers. NASA centers are responsible for providing end users with required facilities and maintaining existing facilities. Headquarters engineering staff monitor project schedules, the obligation of funds, and costing and ensure that project requirements are incorporated as authorized. In-house and consultant architecture-engineering firms develop project plans, designs, drawings, and specifications used by construction contractors. A center's engineering staff oversees construction to ensure that contract requirements, schedules, and funding limits are met. At the center level, an in-house project manager is usually responsible for overseeing a project from start to finish. The project manager contracts for design services, monitors design activities, conducts design reviews, and works with the construction contractor. For a few projects, the design is completed by in-house staff because of the unique nature of the project or for purposes of efficient workforce utilization.

NASA has outsourced management functions for facility acquisition in its government-owned, contractor-operated (GOCO) centers since its establishment. In GOCO centers, the planning, design, construction, and maintenance of facilities is an embedded minor function in support of a center's primary mission, such as research and development or the exploration of space. In a GOCO arrangement, the responsibilities are covered in a cost-plus-award fee contract in which the facilities portion is small compared to the scope of the contract for research and development or other missions. Planning, design, and construction are subcontracted to architecture-engineering and construction firms. The GOCO contractor develops construction projects within allocated resources and in support of the center's mission. Headquarters facilities engineers work with the GOCO contractor's engineers as if they were government employees except that projects are placed under the GOCO contract via contract task order agreements.

To illustrate these procedures, the Advanced Solid Rocket Motor (ASRM) facility was described. The ASRM contract was a cost-plus-award fee contract for the design, development, testing, and manufacture of a rocket. The contract provisions included all testing of manufacturing equipment and facilities. The prime contractor was responsible for generating all facility/equipment criteria and was required to subcontract with a facilitation subcontractor to manage the design and construction activities. The contractor was free to develop the initial design and cost estimate, subject to NASA approval. Once approved, the design was locked into a configuration-control process. The contractor was delegated change authority up to a limited dollar threshold, above which NASA approval was required.

The key results of outsourcing management functions noted in the NASA questionnaire were shorter project delivery time and compensation for staff shortages.

U.S. Department of Energy

DOE's organization is comprised of 12 headquarters program offices, 10 major operations offices, and two large field offices with more than 50 major contractor-operated facilities (NRC, 1999). The facilities infrastructure is managed by a "limited core staff of professional managers and engineers" who primarily oversee the work of a large cadre of contractors working under comprehensive management and operations contracts (NRC, 1999). At DOE, outsourcing of management functions for facilities acquisition is standard operating procedure.

DOE conducts technically complex activities for the federal government, including developing and producing nuclear weapons; operating nuclear reactors, and performing research and development on the military and civilian uses of nuclear energy; promoting and funding nuclear and other sciences; promoting energy conservation and efficiency; managing federal petroleum reserves; and, cleaning up environmental contamination resulting from its past operations (GAO, 1997). These activities often involve large-scale, first-of-a-kind projects that require substantial design, construction, technology, and other expenses. For this study, the committee received briefings on two DOE projects: the Superconducting Super Collider (SSC) Project and the Advanced Photon Source (APS) Project.

Site selection and design activities for the SSC were begun in the 1980s. However, the project was terminated in 1996 prior to construction but after more than $735 million had been spent. The following reasons were cited for the failure of this project:

- Staff was not focussed to build a construction project.
- Management buy-in to approved costs was not demonstrated.
- A committed approach to cost control was not evident.
- The contractor lacked experience in leadership positions.
- Contractor teaming with partners was not effective.
- A trusting relationship between DOE and the contractor did not develop.
- Schedule and planning were not emphasized.
- Control over subcontractors was inadequate.[5]

After evaluating this experience, DOE determined that the following factors would be critical to the success of subsequent projects:

- strong leadership and experience in senior management
- checks and balances at appropriate points
- a strong partnership and shared goals between DOE and the contractor
- a focus on quality
- proactive identification, tracking, and resolution of problems

The APS, in contrast, was determined to be a successful example of the outsourcing of management functions for facility acquisition. The APS, a one-million square foot, third-generation synchrotron radiation facility, was completed ahead of schedule and under budget. The following reasons were cited for this success:

- well defined responsibility and authority for participants
- good conceptual design and cost estimate
- project baseline and change-control system established early

[5]Additional information relevant to the history of the SSC project can be found in *Improving Project Management in the Department of Energy* (NRC, 1999).

- effective communication and periodic reviews
- a common objective
- committed team
- reasoned judgment at every step

In 1997, GAO reviewed 80 DOE projects that cost about $100 million or more and found that the majority of them were behind schedule and over budget. GAO identified the following key factors underlying cost overruns, schedule slippages, and termination of projects (GAO, 1997):

- Constantly changing missions for DOE make it difficult to maintain departmental and congressional support for long-term, high-cost projects.
- Incremental funding of projects from year to year rather than up-front funding, subjects projects to potential delays or terminations.
- A flawed system of incentives both for DOE employees and contractors often rewards contractors despite poor performance.
- Hiring, training, and retaining enough people with the requisite skills to provide effective oversight and/or management of contractors' operations is difficult.

U.S. Air Force

The U.S. Air Force Office of the Civil Engineer, Engineering Division, is responsible for policy, planning, and budgeting of the Air Force's military construction program. The Air Force's FY 96 military construction budget was $587.2 million, including $26.6 million for project design. The program involved 124 projects, including base infrastructure, runways, aircraft hangers, dormitories for enlisted personnel, and child development centers. Almost all planning, design, and construction-related activities of the U.S. Air Force are managed by USACE and NAVFAC, as required by the National Security Act of 1947. In the response to the questionnaire, the Air Force stated that it directly manages less than 5 percent of the total, only those projects for which the Air Force has specific design expertise in a particular type of facility.

NAVFAC and USACE goals are aligned with those of the Office of the Civil Engineer to establish performance measures or other methods of measuring achievements. Those goals are: (1) to award 100 percent of construction projects in the year of budget appropriation and (2) to be ready to award 100 percent of projects anticipated to receive appropriations in the next fiscal year budget. No additional information related to the Air Force's experience outsourcing its management functions was provided.

National Institute of Standards and Technology

At NIST, program/project management is the responsibility of several groups. The Plant Division nominates a five-year plan of active and future projects. Projects for the upcoming fiscal year are evaluated and prioritized based on current and future safety, capacity, and maintenance needs. Upon budget approval, the Facility Planning and Programming Group defines the scope of work for the funded projects. The Design Engineering Group, with the support of private-sector architect-engineer firms, when necessary, produces the final design packages for the Construction Contracts Management Group. Internal end-users work with the Design Engineering Group and architect-engineer firm representatives during design-development.

When the questionnaire was submitted, NIST had outsourced management functions for construction of a large, multimillion-dollar project to a private-sector firm. NIST reported that the reason for the outsourcing was to compensate for a lack of in-house expertise and staff shortages because of a hiring freeze. NIST personnel were involved in all decisions related to changes in project scope and associated costs. Standard procedures included Plant Division Staff acting as the contracting office; a technical representative was assigned to evaluate and validate all change orders for the contracting office. The construction management contract documents defined the duties and responsibilities of the construction management firm; the architect-engineer firm that designed the project and the construction contractor reported to the construction management firm.

NIST reported that a key result of outsourcing the management functions for this project was compensation for a shortage of in-house staff. Outcomes of the project itself were not available because the work was ongoing.

Air National Guard

ANG's physical plant consists of more than 90 military bases located at 160 sites throughout the United States with an aggregate value of approximately $12 billion. The FY 99 annual construction and replacement budget was about $250 million (FFC, 2000).

The National Guard Bureau in Washington, D.C., which provides overall guidance for ANG operations in states, territories, and the District of Columbia during peacetime, also provides design standards and special requirements and monitors the funding, scope, cost, and schedule of major projects. The design process (including design policy, standards, and guidelines) for ANG facilities is managed by the Engineering Center of the Air National Guard Readiness Center (ANGRC/CEC), located at Andrews Air Force Base, Maryland. ANGRC/CEC is also responsible for management oversight and design approval for all major facility project designs. ANGRC/CEC receives technical assistance from one of its branches, the Civil Engineering Technical Services Center, located at Minot, North Dakota.

For new projects, the local engineer at each military base develops the project statement of work based on information provided by the intended facility user or occupant. The base engineer and the facility user or occupant work directly with the design architect-engineer firm to develop the requirements milestones. All design work is contracted out to architect-engineer firms. A federal contracting officer in each state contracts for architect-engineer services, manages construction bidding, and awards construction contracts (FFC, 2000).

In response to the questionnaire, the ANG reported that it had outsourced some management functions for 26 projects in FY 96 to compensate for staff shortages and to shorten project delivery time. All of the projects were outsourced to private-sector firms. The ANG had not yet had enough experience with this outsourcing initiative to report results or outcomes.

Conclusions

Information about the experiences of seven federal owner agencies that outsourced management functions was available to the study committee. Since its establishment, the U.S. Air Force has been required by law to outsource almost all management functions for facility acquisition to USACE and NAVFAC. NASA and DOE, established after the Eisenhower administration, have relied from the beginning on outside contractors to provide many management functions through GOCO arrangements. The remaining four agencies, the U.S. Department of State, IBB, NIST, and ANG, had outsourced management functions to compensate for staff shortages and for the lack of in-house expertise. All seven agencies outsourced management functions for reasons other than cost competitiveness and, therefore, were not required to conduct A-76 analyses.

The Air Force, State Department, and IBB had outsourced management functions to other federal agencies. The Air Force reported that its goals were aligned with those of the provider agencies. The State Department reported that its experience using federal provider agencies was less successful than with private-sector firms because of conflicts with established procedures and internal administrative processes. The IBB reported that its experience with a federal provider agency was successful. Based on this information, the study committee was unable to draw any general conclusions regarding the outsourcing of management functions to other federal agencies.

The State Department and the IBB outsourced management functions for a limited time to compensate for staff shortages and lack of in-house expertise. In both cases, federal staff were able to learn from the external entities and gain enough expertise through training to enable the agencies to resume management functions by in-house staff after the outsourcing contracts had been fulfilled.

NASA and DOE have outsourced management functions through GOCO arrangements since their creation. DOE's experience has been well documented and has generally not been successful for facility acquisition. The reasons for this lack of success are many and varied. At the time the study committee's questionnaire was distributed, the ANG and NIST had outsourced management functions to compensate for staff shortages and a lack of in-house expertise. These projects were ongoing, and the outcomes were not available.

From the available information, the study committee was not able to identify any discernible trends.

SUMMARY

The federal government has contracted out for construction services throughout its history and for design services for more than 70 years. Federal design and construction activities, once the purview of the Treasury Department, USACE, and the Navy Bureau of Yards and Docks, are now managed by at least 25 separate federal entities. Individual agencies may be responsible for managing a facilities program comprised of several dozen to several hundred projects in various stages of planning, design, and construction. Agencies that only require the use of facilities may use a procuring entity to represent the government-as-owner in the acquisition process. Procuring entities include separate executive departments, such as GSA, private-sector firms, and offices within the same agency, such as the FBO within the U.S. Department of State. A few agencies, primarily GSA, USACE, and NAVFAC, provide facilities for other agencies and organizations as a key component of their missions.

The federal government has established general guidance for facilities acquisition through legislation and regulations. The complex, diverse nature of federal projects and the decentralized nature of facilities acquisition, however, preclude an exact, systematic, or single process for programming, budgeting, planning, design, or construction. Within the guidance provided, agencies have developed policies, practices, criteria, and/or guidelines that reflect their missions, cultures, and resources. Studies by academicians and research organizations have concluded that the predesign phases, when decisions are made about the size, function, general character, location, and budget of a facility, are critical to success; the effective commitment of time and resources at this point dictate the future success of the project.

Agencies may use a variety of contract methods to acquire facilities, including design-bid-build, design-build, construction management, and program management. The level of involvement and oversight by the owner organization varies depending on the contract method used.

In the last 50 years, efforts have been made to develop laws, policies, and guidelines for federal agencies to determine which governmental functions should be performed only by government employees (inherently governmental activities) and which can be performed either by government employees or private-sector contractors (commercial activities). Although there is no debate that design and construction activities are commercial, the distinction for management of those functions is not as clear.

The experiences of federal agencies related to the outsourcing of management functions for facility acquisition to other federal agencies and to private-sector firms as reported to the study committee are inconclusive. The most frequently cited reasons for outsourcing management functions were staff shortages and the lack of in-house expertise. The outcomes of outsourcing ranged from failure to success. A number of factors were cited for these outcomes.

FINDINGS

Finding. The outsourcing of management functions for planning, design, and construction services has been practiced by some federal agencies for years. Management functions have been outsourced either to other federal agencies or the private sector. The outcomes of these efforts have varied widely, from failure to success.

Finding. At different times, an agency may fill one or more of the role(s) of owner, user, or provider of facilities.

REFERENCES

BRT (The Business Roundtable). 1997. The Business Stake in Effective Project Systems. Construction Cost Effectiveness Task Force. Washington, D.C.: The Business Roundtable.

Childs, D.M. 1998. Privatization, Outsourcing, and Competition and How It All Relates to the March 1996 OMB Circular A-76. Briefing by David Childs, Office of Management and Budget, to the Committee on Outsourcing of Design and Construction Management-Related Activities for Federal Facilities, National Research Council, Washington, D.C., September 22, 1997.

EOP (Executive Office of the President). 1983. Performance of Commercial Activities. Circular No. A-76. Washington, D.C.: Office of Management and Budget.

FFC (Federal Facilities Council). 2000. Adding Value to the Facility Acquisition Process: Best Practices for Reviewing Facility Designs. Technical Report #139. Washington, D.C.: National Academy Press.

GAO (General Accounting Office). 1992. Government Contractors: Are Service Contractors Performing Inherently Governmental Functions? Report to the Chairman, Federal Service, Post Office and Civil Service Subcommittee, Committee on Governmental Affairs, U.S. Senate. Washington, D.C.: Government Printing Office.

GAO. 1997. Department of Energy: Opportunities to Improve Management of Major Systems Acquisitions. Chapter Report. RCED-97-17. Washington, D.C.: Government Printing Office.

GAO. 1998. OMB Circular A-76. Oversight and Implementation Issues. T-GGD-98-146. Washington, D.C.: Government Printing Office.

OMB (Office of Management and Budget). 1999. Circular No. 1-76 Revised Supplemental Handbook: Performance of Commercial Activities. Washington, D.C.: Office of Management and Budget.

NRC (National Research Council). 1989. Improving the Design Quality of Federal Buildings. Building Research Board, National Research Council. Washington, D.C.: National Academy Press.

NRC. 1990. Achieving Designs to Budget for Federal Facilities. Building Research Board, National Research Council. Washington, D.C.: National Academy Press.

NRC. 1994. Responsibilities of Architects and Engineers and Their Clients in Federal Facilities Development. Board on Infrastructure and the Constructed Environment, National Research Council. Washington, D.C.: National Academy Press.

NRC. 1998. Stewardship of Federal Facilities: A Proactive Strategy for Managing the Nation's Public Assets. Board on Infrastructure and the Constructed Environment, National Research Council. Washington, D.C.: National Academy Press.

NRC. 1999. Improving Project Management in the Department of Energy. Board on Infrastructure and the Constructed Environment, National Research Council. Washington, D.C.: National Academy Press.

PMI (Project Management Institute). 1996. A Guide to the Project Management Body of Knowledge. Sylva, N.C.: PMI Communications.

3

Ownership Functions and Core Competencies

The reliance of federal agencies on nonfederal entities to provide management functions for facility acquisition has raised concerns about the level of control, responsibility, and accountability being transferred to outside entities. Outsourcing of management functions has also raised concerns about the long-term implications for federal agencies' capabilities to plan, guide, oversee, and evaluate facility acquisitions effectively.

Unless a federal agency's mission is to provide facilities, facility acquisition and management is not a core function (i.e., facilities are not the mission but support accomplishment of the mission). However, when acquiring facilities, a federal agency assumes an ownership responsibility as a steward of the public's investment. The expenditure of public funds and the actions undertaken to meet social objectives that may underlie a federal agency's mission require a degree of sensitivity to public issues and concerns that may not be necessary for private-sector organizations. Federal agencies are also responsible for upholding laws and policies that may not apply to the private sector.

One element of the study committee's statement of task was to identify the organizational core competencies federal agencies need for effective oversight of outsourced management functions while protecting the federal interest. This chapter begins with a brief review of the differences between ownership and management functions for facility acquisitions, including a discussion of the relationship between ownership and management functions and inherently governmental functions. The next section describes the characteristics of a smart owner of facilities. Ownership functions and core competencies for owners, users, and providers of facilities are then identified. The chapter concludes with recommendations for the development and retention of core competencies for federal facility acquisitions.

OWNERSHIP AND MANAGEMENT FUNCTIONS

The nature of ownership and management functions in the facility acquisition process differ. Ownership functions are to establish objectives and to make decisions. Management functions, in contrast, include performing ministerial tasks to carry out or implement the owner's decisions and, by definition, to control the accomplishment of the task. Owner functions include determining the need for a facility, developing the project scope, balancing conflicting priorities, establishing parameters (e.g., cost and duration), and determining positions in disputes. Management functions include obtaining information from the owner, contractors, and others; analyzing the information, making recommendations, and determining options; and ensuring that communications are maintained among all parties. Owner and management functions are equally important for successful facility acquisitions. A well defined project led by an owner with a clear vision but with a poor management structure will probably fail. A poorly defined project with a good management structure will also fail but for different reasons.

Federal legislation and policies related to determining which functions are inherently governmental are a critical determinant in deciding which functions can and cannot be outsourced. Section 7.5 of the FAR provides guidance to ensure that inherently governmental functions are not performed by contractors (see Appendix C for the complete text of FAR Section 7.5). In preparing this report, the committee reviewed this regulation and concluded that examples of inherently governmental functions that apply to facilities acquisition are listed in Sections 6, 7, 11, 12, and 16, which are reprinted below:

(6) The determination of federal program priorities for budget requests.
(7) The direction and control of federal employees.
(11) The determination of what government property is to be disposed of and on what terms.
(12) In Federal procurement activities with respect to prime contracts:
  (i) Determining what supplies or services are to be acquired by the Government (although an agency may give contractors authority to acquire supplies at prices within specified ranges and subject to other reasonable conditions deemed appropriate by the agency);
  (ii) Participating as a voting member on any source selection boards;
  (iii) Approving any contractual documents, to include documents defining requirements, incentive plans, and evaluation criteria;
  (iv) Awarding contracts;
  (v) Administering contracts (including ordering changes in contract performance or contract quantities, taking action based on evaluations of contractor performance, and accepting or rejecting contractor products or services);
  (vi) Terminating contracts;

(vii) Determining whether contract costs are reasonable, allocable, and allowable; and

(viii) Participating as a voting member on performance evaluation boards.

(16) The determination of budget policy, guidance and strategy.

Inherently governmental functions include making a decision (or casting a vote) pertaining to policy, prime contracts, or the commitment of government funds. None of the functions listed is ministerial or informational, with the possible exception of "administering contracts." Although the meaning of the term administer may be broad enough to include ministerial and information tasks, the examples of administration listed in Section 12(v) are limited to making decisions on issues likely to arise under the contracts. In essence, therefore, the distinction between inherently governmental functions and commercial activities is the same as the distinction between ownership and management functions.

FAR Section 7.503(d) includes a list of activities that are not ordinarily considered inherently governmental but that may result in a contractor acquiring knowledge or wielding influence ordinarily considered more appropriate for an owner than a manager. Activities that relate to the acquisition of facilities include:

(3) Services that involve or relate to analyses, feasibility studies, and strategy options to be used by agency personnel in developing policy.

(5) Services that involve or relate to the evaluation of another contractor's performance.

(6) Services in support of acquisition planning.

(7) Contractors providing assistance in contract management (such as where the contractor might influence official evaluations of other contractors).

(8) Contractors providing technical evaluation of contract proposals.

(9) Contractors providing assistance in the development of statements of work.

(11) Contractors working in any situation that permits or might permit them to gain access to confidential business information and/or any other sensitive information (other than situations covered by the National Industrial Security Program described in 4.402(b)).

(14) Contractors participating as technical advisors to a source selection board or participating as voting or nonvoting members of a source evaluation board.

(15) Contractors serving as arbitrators or providing alternative methods of dispute resolution.

(16) Contractors constructing buildings or structures intended to be secure from electronic eavesdropping or other penetration by foreign governments.

(17) Contractors providing inspection services.
(18) Contractors providing legal advice and interpretations of regulations and statutes to Government officials.

Although the activities listed above are not inherently governmental functions, the FAR cautions agencies to consider whether a contractor's performance of them might unduly impinge on the ownership function of decision making and confidentiality. Section 12(e) explains that the agency's decision about outsourcing these activities depends on the degree to which ownership functions may be compromised:

> This assessment should place emphasis on the degree to which conditions and facts restrict the discretionary authority, decision-making responsibility, or accountability of Government officials using contractor services or work products.

In the committee's opinion, FAR Section 7.5 can be used as a the basis for a two-step threshold test for determining whether a particular management function related to facilities acquisition should be performed by federal agency staff to protect the federal interest. The first step is to determine whether the function involves decision making on important issues (ownership) or involves ministerial or information-related services (management). If it is an ownership function, it should be performed by in-house staff and should not be outsourced. If it is a management function, the second step of the analysis is to consider whether the function might unduly compromise one or more of the agency's ownership functions, particularly those listed in Section 3(d). If it would, then the function should be considered a "quasi"-inherently governmental function and should not be outsourced.

Figure 3-1 shows how functions can be grouped for decision-making purposes. Functions that fall into the category of "internal-dedicated" would be judged too critical to outsource. At the other extreme, functions identified as "external-shared" could be outsourced with relatively little concern. Functions that fall into the categories of "internal-shared" and "external-dedicated" require additional analysis to determine if outsourcing is appropriate. The Army's experience with the Logistics Civilian Augmentation Program, known as LOGCAP, is a case in point. A private contractor has provided almost all facility services, including project management, for six of the Army's recent deployments. This outsourcing of an external-dedicated function shows that even activities intimately connected with an agency's core mission can sometimes be outsourced successfully.

If a function survives this threshold analysis and is deemed to be a management function that does not unduly compromise the agency's ownership functions, then the agency should determine whether or not to outsource the activity, based on a number of factors (these are addressed in Chapter 4). Agencies should be wary of compromising their ownership responsibilities and functions. The line

	Internal resources	*External resources*
Dedicated	Functions critical to the accomplishment of the core mission	Functions that can be performed by others but are full time and dedicated to the accomplishment of the core mission
Shared	Functions that support the core mission of the agency and need to be performed with internal resources	Functions that can be shared with other agencies or outside resources

FIGURE 3-1 A four-square analysis tool to determine whether functions could be outsourced.

between inherently governmental functions and commercial activities and between ownership and management functions can be very fine, and distinguishing between them can be difficult. Therefore, projects should be analyzed on a case-by-case basis.

CHARACTERISTICS OF A SMART OWNER

The term *smart owner* is used in the commercial design, engineering, and construction industry to designate a business entity that has the skill base—usually a staff with the professional qualifications and authority—necessary to plan, guide, and evaluate the facility acquisition process. A smart owner focuses on the relationship of a specific facility to the successful accomplishment of an organization's business or overall mission.

A smart owner of facilities must be capable of performing four interdependent functions related to acquisition (Figure 3-2):

- establishing a clear project definition
- establishing progress metrics

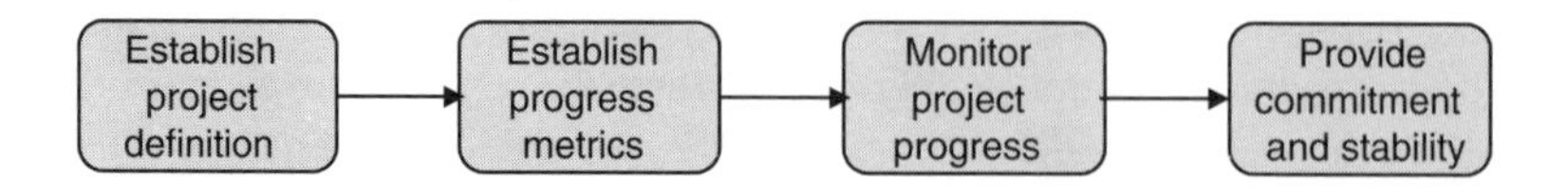

FIGURE 3-2 The four owner functions in successful facility acquisition.

- monitoring overall project progress
- providing commitment and stability to the project definition and its achievement (i.e., leadership)

Establishing Project Definition

The first ownership function in the successful acquisition of a facility is the establishment of a clear project definition. Industry research has shown that preproject planning (requirements assessment, conceptual planning, and programming phases) has the greatest potential impact on project outcome. Thus, even if an organization outsources management functions for planning, design, and construction activities, competent representatives of the owner must still lead and implement preproject planning. Setting project-specific goals, objectives, and priorities requires knowledge of the organization's overall business or mission and the ability to translate facility requirements to meet business or mission objectives (FFC, 1998). Specific tasks may include:

- developing a strategic plan or written scope statement that defines mission needs, relates them to project requirements, and serves as the basis for future project decisions and control of changes in scope
- preparing an integrated project plan that addresses the overall strategy for acquiring the end product and/or services and identifies interfaces, including regulatory interface points and requirements
- preparing a detailed execution plan and schedule to establish the tactics, organizational relationships, roles, and responsibilities for accomplishing various aspects of the project (NRC, 1999)

The smart-owner function becomes increasingly important when organizations, including federal agencies, use design-build and project management contract methods that limit owner involvement after the preproject planning phases.

Establishing Progress Metrics

The second smart-owner function, the establishment of progress metrics, requires that project objectives be translated into measurable criteria. The criteria should include not only absolute constraints (e.g., allocated funding, delivery schedule, performance specifications), but also the relative rate of progress that reveals the probability of completion within the constraints. For example, the metrics should include the rate of financial expenditures in different cost categories (e.g., labor, materials, and equipment) expected to be required to complete the project on schedule. Although the collection of data and the measurement of progress can be outsourced, the development of the metrics should be the responsibility of the owner organization to ensure that they incorporate critical program directives and are commensurate with the organization's mission.

Monitoring Project Progress

A smart-owner organization uses detailed data collected and aggregated from the field to monitor progress. A key objective of monitoring is the active identification and mitigation of project (and program) risks (i.e., determining which risks are likely to affect the project and documenting the characteristics of each). Although the actual monitoring can be outsourced, the overall assessment of project performance should be conducted by the owner organization. Decisions related to risk identification and mitigation should also be the owner's responsibility.

Providing Commitment, Stability, and Leadership

The owner organization is responsible for the successful completion of a project and, therefore, has the authority to commit resources for that project and ensure stability throughout its duration. Project stability requires that progress related to specific metrics defined early in the life of the project be continuously monitored and maintained. Vacillations on performance objectives (e.g., allowing cost overruns to occur routinely) can be fatal to the successful acquisition of a facility.

In performing these functions, the owner organization is responsible, by definition, for providing leadership, which involves the following responsibilities (PMI, 1996):

- establishing direction (developing both a vision of the future and strategies for changes to achieve that vision)
- aligning people (communicating the vision by words and deeds to all those whose cooperation may be needed to achieve the vision)
- motivating and inspiring (helping people energize themselves to overcome political, bureaucratic, and resource barriers to change)

GAO has also recognized the importance of a committed senior leadership team in fulfilling an agency's mission and in establishing a vision for the future, core values, goals, and strategies. According to GAO, essential functions of senior leadership are "aligning organizational components so that the agency can best pursue this vision and building a commitment to the vision at all levels of the organization" (GAO, 1999a).

Although activities related to the management of specific projects can be outsourced, the owner organization is ultimately accountable for the performance, cost, quality, and functionality of the acquired facility. Therefore, the owner organization should provide leadership and retain in house the functions that are the most significant determinants of success. To do this, the owner organization must retain in-house staff with the management, financial, communication, and technical skills necessary for effective oversight of the acquisition process, from scope definition to start-up of the facility.

CORE COMPETENCIES FOR FACILITY ACQUISITIONS

In reviewing the characteristics of the best facility acquisition systems, The Business Roundtable found that owner organizations with better-than-average project acquisition systems all maintained some form of central facilities engineering organization, which was responsible for "providing excellence in project definition, maintaining disciplinary excellence in project management...[and] integrating contractors effectively into their project process." Those same skills helped the organizations "select the right capital assets to make, acquire, or refurbish" (BRT, 1997). A recent NRC report found that the "best public agencies and private firms engaged in capital project development maintain central organizations with core competencies in project management, project planning, coordination, and human resources development." Such organizations provide "structure, continuity, and leadership that foster cooperation both internally and externally" (NRC, 1999). And the Center for Construction Industry Studies has found that "using project teams and retaining in-house expertise in key functional areas of engineering improves the owner's ability to control project outcomes, evaluate contractor performance, and make informed decisions about contractor selection. Retaining this expertise in-house means that the owner is not dependent on just one person for the success of a project" (CCIS, 1999).

Core competencies constitute an organization's essential area of expertise and skill base. In *Competing for the Future* (Hamel and Prahalad, 1994), a competence is defined as a:

> bundle of skills and technologies rather than a single discrete skill or technology....A core competence represents the sum of learning across individual skill sets and individual organizational units. Thus, a core competence is very unlikely to reside in its entirety in a single individual or small team.

Unless a federal agency's mission is to provide facilities, facility acquisition and management are not core functions because facilities are not the agency's mission but support the accomplishment of the mission. However, as a steward of the public's investment in facilities, federal agencies have a responsibility to be smart owners.

A federal agency's responsibility to be accountable for upholding public policy and its authority to commit public resources are indivisible. This combination of responsibilities requires that any federal agency acquiring facilities have the in-house capabilities to perform the owner's functions at the top administrative levels. The agency must have the capacity to translate its mission needs directly into program definitions and project specifics and otherwise act in a publicly responsive and accountable manner. However, other organizational core competencies needed to direct and manage specific projects effectively vary, depending on whether the agency is acting as an owner, user, or provider of facilities.

Core Competencies for Owners

To function effectively as an owner when acquiring facilities, a federal agency should have the organizational core competencies to perform the following functions:

- evaluate and implement government-wide and agency-specific policies and standards and suggest ways to improve them
- develop, analyze, select, implement, and adjust the means or alternatives to achieve program or project objectives
- monitor, control, and adjust program or project implementation based on specific progress metrics (e.g., cost, schedule, complete-to-date, cost-to-complete)

To manage planning, design, and construction services effectively or to oversee the management of those services by an outsourced entity, the agency should also have the capabilities to perform the following functions:

- detailed technical analyses of alternatives, including design, procurement, construction, and final performance requirements
- financial analyses of the relative costs, benefits, and cash flows of the alternatives from conceptualization through design, procurement, and construction to start-up
- project management analyses for the identification, collection, analysis, and summary of accurate and valid project data
- construction-management activities to implement and adjust policies, standards, and resource allocations to project conditions

In short, federal agencies should retain the core competencies to establish project definitions, establish project metrics, monitor project progress, and ensure commitment, stability, and leadership. Owner agencies should have the leadership capability to develop and drive the process to increase the probability of success. Therefore, they should maintain in-house staffs capable of performing financial and technical analyses, as well as providing project and general management. In business terms, "critical owner skills include technical knowledge of the process, alignment with the business units' goals and objectives, facility definition, stewardship of the overall project process and objectives and project controls" (Sloan Program for the Construction Industry, 1998).

TABLE 3-1 Skills Required by Successful Owner Project Personnel

Category of Skills	Examples of Skills
Business	Writing and managing contracts
	Negotiation
	Managing budgets and schedules
Communication	Coordination/liaison
	Conflict management
	Cultivate broad network of relationships
Influence	Mentoring
	Motivating
	Change management
Managerial	Team building
	Delegating
	Politically aware/see big picture
Problem Solving	Continually analyze options/innovation
	Planning
	Consider both sides of issues, risk management
Technical	Understand entire construction process
	Multidisciplined (knowledge of several areas of engineering)
	Information technology skills

Source: CCIS, 1999.

The Center for Construction Industry Studies has identified a variety of specific skills related to the core competencies necessary for smart owners when outside entities are used extensively. These skills have been grouped into six categories (see Table 3-1).

When some functions are outsourced, project-management skills become vitally important for owner organizations. Project management has been defined as " the application of knowledge, skills, tools, and techniques to project activities in order to meet or exceed stakeholder needs and expectations from a project...[it] invariably involves balancing competing demands among scope, time, cost and quality; stakeholders with differing needs and expectations; identified requirements (needs) and unidentified requirements (expectations)" (PMI, 1996). One essential condition for successful facility acquisition identified in a recent NRC report (1999) was:

> Project managers (in owners' as well as contractors' organizations) are experienced professionals dedicated to the success of the project. Each demonstrates leadership, is a project team builder as well as a project builder, possesses the requisite technical, managerial, and communications skills, and is brought into the project early.

Core Competencies for Facility Users

If an agency's role is that of facility user, rather than owner, the agency is responsible for acting as a "smart buyer" of services, including design, construction, and management services. To be a smart buyer, an agency should retain in-house personnel who understand the agency's mission and requirements, as well as customer needs, and who can translate those needs and requirements into the agency's strategic direction (FFC, 1998). For a project to be successful, the project sponsors must "know what they need and can afford, where they want to locate the project and when it must be ready for use or otherwise completed." Facility users should be committed to project scope, requirements, budget, and schedule and should have the capacity to weigh options and make timely, informed decisions to avoid project delays (NRC, 1999).

Core Competencies for Providers of Facilities

Agencies or entities whose mission includes providing facilities have a greater need to retain technical, general, and project management core competencies to ensure that they provide quality facilities that meet owner and user agencies' needs. General management "encompasses planning, organizing, staffing, executing and controlling operations of an on-going enterprise" (PMI, 1996). General management skills include leading (as defined above), communicating (verbally and in writing), negotiating, problem solving, influencing the organization, and the ability to get things done (PMI, 1996).

NAVFAC provides facility engineering services to all of the Navy and Marine Corps, to other U.S. Department of Defense services and agencies, as directed, and to federal agencies and others on a case-by-case basis. NAVFAC has four engineering field divisions, each of which provides a full range of construction services, including project management, contracting functions, and construction management. For most construction projects, NAVFAC also manages the design phase; about 10 to 15 percent of the designs are accomplished with NAVFAC staff (FFC, 2000). In response to the committee's questionnaire, NAVFAC identified its core competencies as master planning, project planning, cost estimating, engineering, design, construction, acquisition, and project management. NAVFAC also considers an understanding of the Navy's mission, standards, and procedures a core competency.

USACE defines its core competencies as "a set of interwoven skills tied to information systems and organizational values, a complex set of skills, capabilities and expertise that reside in employees working within and across skill sets" (FFC, 1998). Identified core competencies include the following capabilities:

- respond quickly through its worldwide organization
- quickly and effectively staff up to any size project with in-house and external resources
- provide a structured, rational approach to problem solving and a process for "best fit" solutions

- facilitate or broker cooperative arrangements for public and private constituents
- offer full life-cycle project services
- implement public policy within the Army ethic

USACE also identified specialized engineering services and project management as specific core competencies (FFC, 1998).

The other major provider of federal facilities, GSA, defines its mission as "provid[ing] expertly managed space, products, services and solutions at the best value and policy leadership to enable federal employees to accomplish their missions" (FFC, 1998). To fulfill its mission, GSA has reorganized itself as a portfolio-management organization with four primary goals: to promote responsible asset management; to compete effectively for the federal market; to excel at customer service; and to anticipate future workplace needs. To meet these goals, GSA project managers must have the management and technical skills to (FFC, 1998):

- align resources to the workload
- use technology effectively
- organize staffs and lead them towards a common goal of delivering a project on time and within budget
- manage plans, schedules, budgets, expenditures and change orders

Development and Retention of Core Competencies

Federal agencies should retain the organizational core competencies necessary to act as smart owners or smart buyers when acquiring facilities. Provider agencies require additional technical competencies in engineering, architecture, general management, and project management to perform effectively. Because agencies' roles in acquisition vary, the types of federal facilities acquired also vary widely. In addition, a wide range of new and evolving contract methods for project delivery have inherently different levels of risk and management requirements. For these reasons, no single approach or set of core competencies for the acquisition of federal facilities can be applied to all agencies or situations. Senior leaders and staff of each agency should identify the organizational core competencies necessary for effective facility acquisitions to support their current and future missions.

Federal agencies face a number of challenges in developing and retaining core competencies for facility acquisitions. As part of its performance and accountability series, GAO issued a series of reports on the major management challenges and program risks facing federal agencies. Among its findings were: (1) the federal government's performance has been limited by a failure to manage projects on the basis of a clear understanding of the results that agencies are to achieve and how performance will be gauged; (2) major challenges must be overcome, both at the agency level and for the U.S. government as a whole, in preparing reliable financial statements; and (3) human-capital planning must be an integral part of an organization's strategic and program planning (GAO, 1999b). The report goes on to note that, because of the rapid pace of social and technological change, shifts in agency missions and strategies to achieve their missions, combined with downsizing, agencies are "continually faced with the challenge of attracting, retaining, and motivating appropriately skilled staff." As a consequence, "skills gaps in critical mission areas undermine agencies' effectiveness and efforts" (GAO, 1999b). DOE, for example, has reported that the "lack of skilled staff in program and contracting oversight positions is one of the most fundamental challenges for the department" (GAO, 1999b). The ability of the U.S. Department of Housing and Urban Development to perform essential functions, such as monitoring multibillion-dollar programs, has been limited by "not having enough staff with the necessary skills." Of the 32,000 financial management personnel employed by the U.S. Department of Defense, less than half were given any financial or accounting-related training in 1995 or 1996, a time when the department was attempting to implement significant accounting reforms (GAO, 1999b).

Problems in attracting, training, and retaining qualified staff for facility acquisition are not confined to the federal sector. A report by the Center for Construction Industry Studies based on 274 projects from 31 public and private-sector organizations showed that approximately 62 percent of planning, design, and procurement functions were outsourced. Based on detailed case studies of three of these organizations and interviews with members of 22 other organizations, the study found that it is "fairly well recognized in owner firms that the skill set required to manage and work on projects from the owner's side has changed dramatically...[and] the issue of skill development of owner personnel is perhaps the most important difficulty facing owner firms" (CCIS, 1999). The surveyed firms had "invested relatively little systematic effort into methods for ensuring that their personnel have the required skill sets" or formal training. Instead, they relied on on-the-job training for new employees and on the "few experienced personnel they have retained in-house" (CCIS, 1999). The report noted that, as the "current cadre of long-tenured individuals retires and need to be replaced, the effects of lack of training will become more critical" (CCIS, 1999).

At most federal agencies, the major portion of operating costs is devoted to personnel costs and salaries. For this reason, employees have "often been seen as costs to be cut rather than assets to be appreciated" (GAO, 1999b). However, business management research has shown the need for continual "organizational learning" to improve the effectiveness and efficiency of organizational functions. "High performance organizations in both the public and private sectors recognize that an organization's people largely determine its capacity to perform" (GAO, 1999b):

> ...a high-performance organization demands a dynamic, results-oriented workforce with the talents, multidisciplinary knowledge, and up-to-date skills to enhance the agency's value to its clients and ensure it's equipped to achieve its mission. Because mission requirements, client demands, technologies, and other environmental influences change rapidly, a performance-based agency must continually monitor its talent needs...In addition, this talent must be continuously developed through education, training, and opportunities for continued growth.

Federal agency staffs need a broad range of management, technical, communication, and leadership skills to act as effective stewards when acquiring facilities for the government. Agency leaders should evaluate current organizational skills, identify organizational skills likely to be lost through attrition, retirement, or continued reductions, forecast needs based on projected workloads, technologies, and contract types. A number of approaches can then be used to acquire, develop, and retain the necessary organizational core competencies and skills. Each agency will have to determine which approach or combination of approaches will be the most effective for its specific circumstances.

One approach is to hire personnel from the public or private sector who have the training and experience necessary to perform these functions. A second approach is to provide the training and professional development for in-house staff to acquire necessary skills.

Project management is increasingly being recognized as a professional discipline. The Project Management Institute, the Association for Project Management, the Australian Institute of Project Management, the Construction Management Association of America, and the International Project Management Association, among others, have developed certification programs for project managers. A recent NRC report found that to satisfy the basic core competencies required for a federal agency to be a smart owner, and for agencies that elect to retain their management activities, the staff involved with implementing capital programs should be trained and certified in project management. This professional training should be updated throughout their federal employment (NRC, 1999).

Agencies can also design and conduct training programs based on industry best practices but tailored to the federal environment. NASA, for example, has developed a training program based on best practices identified by the Construction Industry Institute. Although NASA staff receive first priority for training, staff from other agencies also attend NASA's training course. Agencies should investigate the training and education available by other agencies and by outside organizations before developing their own training programs.

For owner agencies or entities involved in providing facilities, one way to maintain and enhance technical proficiency is by retaining a portion of the planning, design, or construction management in house as part of a professional development program. Junior staffers need "hands-on" experience to develop and enhance their managerial and executive skills. Simply having in-house resources dedicated to a function, however, does not guarantee that technical proficiency will be maintained or enhanced. Professional development programs should be appropriate to the staff level of experience.

Mentoring programs can be an effective approach to on-the-job training, as well as to capturing institutional knowledge. GSA, for instance, has plans to create a learning center where less experienced project managers will have access to information and training. A mentoring program is also planned to encourage people who might be considering retirement to stay on and become mentors to less experienced personnel (FFC, 1998).

Staff training should also focus on acquisition of competencies tailored to reflect an agency's context and requirements. This training should be comparable to the training available to employees of commercial architecture-engineering and construction-management firms. By maintaining professional skills at a level comparable to the skills typical of commercial design and construction firms, training and certification programs can provide a significant incentive for qualified personnel to enter and remain in government service.

Professional development should also be nurtured through tangible and intangible rewards for effective program and project management, including emphasis on leadership and the opportunity to exercise it, management of a portfolio of projects, and the opportunity to advance an agency's strategic objectives through the implementation of specific projects.

SUMMARY

Ownership and management functions in the facility acquisition process differ. An owner's role is to establish objectives and make decisions. Management functions include the ministerial tasks necessary to carry out or implement the owner's decisions. In reviewing Section 7.5 of the FAR, the committee concluded that inherently governmental functions as they relate to facility acquisition involve making a decision (or casting a vote) pertaining to policy, prime contracts, or the commitment of funds and do not include ministerial functions. In essence, therefore, the distinction between inherently governmental functions and commercial activities is the same as the distinction between ownership functions and management functions.

Using Section 7.5 of the FAR as a basis, the committee developed a two-step threshold test for determining whether a particular management function related to facility acquisitions should be performed by federal agency staff to protect the federal interest. The first step is to determine whether the function involves decision making on important issues (ownership) or ministerial or information-related services (management). If it is an ownership function, it should be performed by in-house staff and should not be outsourced.

If it is a management function, the second step of the analysis is to consider whether the function might unduly compromise one or more of the agency's ownership functions. If it would, then the function should be considered a "quasi"-inherently governmental function and should not be outsourced. If a management function survives this threshold analysis, then the agency should determine whether or not to outsource the function based on a number of factors outlined in Chapter 4.

Core competencies constitute an organization's essential area of expertise and skill base. Unless a federal agency's mission is to provide facilities, facility acquisition and management are not core functions (i.e., providing facilities supports accomplishment of the mission but is not the primary goal). However, when acquiring facilities, federal agencies assume an ownership responsibility as a steward of the public's investment. The requirements that a federal agency be accountable for upholding public policy and have the authority to commit public resources are indivisible. This combination of responsibilities requires that any federal agency acquiring facilities have the in-house capabilities to translate its mission needs directly into program definitions and project specifics and otherwise act in a publicly responsive and accountable manner. Other organizational core competencies needed to direct and manage specific projects vary, depending on the agency's role as owner, user, or provider of facilities.

A smart owner of facilities must be capable of performing four interdependent functions related to acquisition: define project scope, goals, and objectives clearly; establish performance criteria to evaluate success; monitor project progress; and provide commitment and stability, (i.e., leadership) for achieving the goals and objectives.

FINDINGS AND RECOMMENDATIONS

Finding. Each federal agency involved in acquiring facilities is accountable to the U.S. government and its citizens. Each agency is responsible for managing its facilities projects and programs effectively. Responsibility for stewardship cannot be outsourced.

Finding. Key factors in determining successful outcomes of outsourcing decisions include clear definitions of the scope and objectives of the services required at the beginning of the acquisition process and equally clear definitions of the roles and responsibilities of the agency. Owners and users need to provide leadership; define scope, goals, and objectives; establish performance criteria for evaluating success; allocate resources; and provide commitment and stability for achieving the goals and objectives.

Finding. Program scope, definition, and budget decisions are inherently the responsibilities of owners/users and should not be outsourced. However, assistance in discharging these responsibilities may have to be obtained by contracting for services from other federal agencies or the private sector.

Finding. The successful outsourcing of management functions by federal agencies requires competent in-house staff with a broad range of technical, financial, procurement, and management skills and a clear understanding of the agency's mission and strategic objectives.

Finding. Because federal facilities vary widely, and because a wide range of new and evolving project delivery systems have inherently different levels of risk and management requirements, no single approach or set of organizational core competencies for the acquisition of federal facilities applies to all agencies or situations.

Finding. The organizational core competencies necessary to oversee the outsourcing of management functions for projects and/or programs need to be actively nurtured over the long term by providing opportunities for staff to obtain direct experience and training in the area of competence. The necessary skills will, in part, be determined by the role(s) the agency fills as owner, user, and/or provider of facilities.

Recommendation. Federal agencies should first determine their role(s) as owners, users, and/or providers of facilities and then determine the core competencies required to effectively fulfill these role(s) in overseeing the outsourcing of management functions for planning, design, and construction services.

Recommendation. Owner/user agencies should retain a sufficient level of technical and managerial competency in house to act as informed owners and/or users when management functions for planning, design, and construction are outsourced.

Recommendation. Provider agencies should retain a sufficient level of planning, design, and construction management activity in house to ensure that they can act as competent providers of planning, design, and construction management services.

Recommendation. Agencies should provide training for leaders and staff responsible for technical, procurement, financial, business, and managerial functions so that they can oversee the outsourcing of management functions for planning, design, and construction services effectively.

REFERENCES

BRT (The Business Roundtable). 1997. The Business Stake in Effective Project Systems. Washington, D.C.: The Business Roundtable, Construction Cost Effectiveness Task Force.

CCIS (Center for Construction Industry Studies). 1999. Owner / Contractor Organizational Changes. Phase II Report. Austin, Texas: University of Texas Press.

FFC (Federal Facilities Council). 1998. Government/Industry Forum on Capital Facilities and Core Competencies. Technical Report #136. Washington, D.C.: National Academy Press.

FFC. 2000. Adding Value to the Facility Acquisition Process: Best Practices for Reviewing Facility Designs. Washington, D.C.: National Academy Press.

GAO (General Accounting Office). 1999a. Human Capital: A Self-Assessment Checklist for Agency Leaders. GGD-99-179. Washington, D.C.: Government Printing Office.

GAO. 1999b. Major Management Challenges and Program Risks: A Government-wide Perspective. Letter Report. OCG-99-1. Washington, D.C.: Government Printing Office.

Hamel, G., and C.K. Prahalad. 1994. Competing for the Future. Boston, Mass.: Harvard Business School Press.

NRC (National Research Council). 1999. Improving Project Management in the Department of Energy. Board on Infrastructure and the Constructed Environment, National Research Council. Washington, D.C.: National Academy Press.

PMI (Project Management Institute). 1996. A Guide to the Project Management Body of Knowledge. Sylva, N.C.: PMI Communications.

Sloan Program for the Construction Industry. 1998. Owner / Contractor Organizational Changes. Phase I Report. Austin, Texas: University of Texas Press.

4

Decision Framework

This chapter provides a decision framework for federal agencies considering outsourcing management functions for facility acquisitions. Based on the constraints of inherently governmental functions, the framework incorporates the committee's two-step threshold for identifying ownership functions, which should be performed by in-house staff, and management functions, which can be considered for outsourcing. The decision framework is not intended to generate definitive recommendations for which management functions may or may not be outsourced or in what combination. It is a tool to assist decision makers in analyzing their organizational strengths and weaknesses, assessing risk in specific areas based on the stature and sensitivity of a project, and, at a fundamental level, questioning whether or not a management function can best be performed by in-house staff or by an external organization.

The outsourcing of management functions for planning, design, and construction activities by federal agencies is a strategic decision that should be considered in the context of an agency's long-term mission. Federal agencies should analyze the relationship of outsourcing decisions to the accomplishment of their missions before outsourcing management functions for planning, design, or construction activities. Outsourcing for services and functions should be integrated into an overall strategy to achieve the agency's mission, to manage resources, and to achieve best value or best performance for the resources expended. Outsourcing of management functions should not be implemented solely as a short-term expedient to limit spending or to reduce in-house personnel.

The outsourcing decision framework comprises the following five steps:

- Can the function legally be outsourced, or is it an inherently governmental function?
- Is the function an ownership or management function?
- Is it wise to outsource—do the characteristics of the project require that this function be managed by agency personnel?
- Is there a need for or advantage to outsourcing—does the agency have the capabilities to perform the management function effectively?
- If outsourcing is not precluded by other factors, is it an appropriate way to proceed?

These steps are illustrated in Figure 4-1.

LEGALITY OF OUTSOURCING

The first step in the decision process is to identify the program, project, or service being considered for outsourcing and its associated management functions. If the function is determined to be an inherently governmental function, as defined in the FAIR Act of 1998 and FAR Section 7.5, then federal employees must perform it. The function could be outsourced to another federal agency but not to a private-sector entity. Organization, staff capability, resources, schedule, and other issues must still be addressed as outlined below.

NATURE OF FUNCTION

An agency is ultimately responsible for the performance of the program, service, or function to be outsourced and, thus, must consider a number of factors before making a decision to outsource it to another federal agency, outsource it to a private contractor, or manage it in house. At this point, the agency should begin to apply the two-step threshold test for determining whether a particular management function should be performed by in-house staff or an external organization. From this point on, federal agencies that provide management functions and private-sector organizations are both considered external organizations. In other words, the decision-making process for outsourcing to a public organization or a private organization is the same.

The first step is to determine whether the function to be outsourced is an ownership function (i.e., one that involves decision making on important issues) or a management function (i.e., one that involves ministerial or information-related services) as described in Chapter 3. In the committee's opinion, if it is an ownership function, it should not be outsourced. Agencies should guard against losing control of their ownership functions and should retain the core capabilities necessary to carry out ownership responsibilities associated with facility acquisitions in their role(s) as owners, users, or providers of facilities.

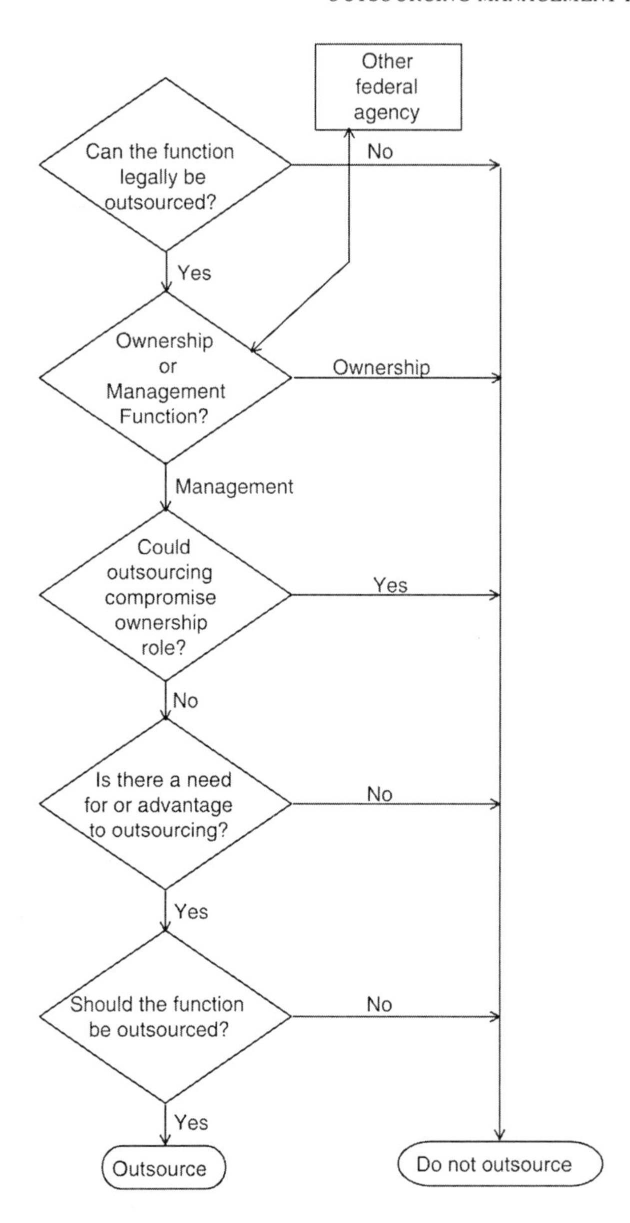

FIGURE 4-1 Decision framework for outsourcing management functions.

WISDOM OF OUTSOURCING

The second step in the threshold test is to consider whether outsourcing a management function might unduly compromise one or more ownership functions, such as restricting the agency's ability to make decisions or compromising its possession of confidential information. A number of project-related management functions should be retained by smart owners and smart buyers (perhaps assisted by contract staff or specialized consultants). These functions relate primarily to strategic issues, such as mission, scope, priority, and budget. In the committee's opinion, if outsourcing the management function would unduly compromise the agency's ownership functions, then the function should not be outsourced.

The following questions should be asked in reaching a decision about outsourcing a management function:

- Will decisions or tasks related to the function have extraordinary / critical results for the success of the project?
- Is the management function one that requires significant fiduciary responsibility that can have an impact on the progress of the project if it is mismanaged?
- If outsourced, will the management function bind the agency to either a monetary commitment or a contract without protecting the public interest?
- Will decisions and tasks related to the function have effects beyond the scope of the project (e.g., environmental, public safety, or national security effects)?
- Will decisions and tasks related to the function infringe on mandates by government or requirements by law?
- If outsourced, will the management function place unjustified and uncontrollable authority in the hands of one private provider over another?
- If outsourced, will the management function require the external organization to make the service delivery sufficiently proprietary to the point that the agency would be committed solely to that organization for future services?

Although an affirmative answer to any one of these questions may not preclude a decision to outsource, each question represents an opportunity for the agency to retain or relinquish control of a critical management function. Because the ultimate responsibility for the program, project, or service remains with the agency, affirmative responses should be carefully and individually considered.

NEED FOR OR ADVANTAGE OF OUTSOURCING

If a management function is deemed to be one that does not unduly compromise the agency's ownership functions, the next step in the decision process is for the agency to determine if there is a need for or advantage to outsourcing. The answer will be partly based on the agency's role as an owner, user, or provider of facilities and its core competencies. A key factor in this determination will be the availability of qualified staff to perform the management function and/or the time available to successfully accomplish the function in house.

Perhaps the most fundamental consideration is whether an agency has in-house professional or technical skills available to manage planning, design, or construction activities. If the activity will not be a continuing one, there would be little point to hiring and training staff to handle a one-time special need or peak workload. If special study skills, or even data systems, are required for the management function, it may not be feasible or cost effective to acquire the skills or technology and train in-house staff to perform the function. Outsourcing in these situations can be an effective or advantageous way of handling unique workloads for which it would be impractical to retain or train in-house staff or purchase equipment.

In a related situation, an agency may have a modicum of in-house competencies but not enough qualified people to handle peak or shifting workloads effectively at a given time. Or an agency may have the required competencies in some field offices but not in others. In these situations, it may be advantageous for the agency to outsource some of its functions to meet project delivery deadlines.

The preceding discussion and the analysis of the federal experience in Chapter 2 demonstrate that the decision to outsource a management function is seldom clear cut. Agencies should consider if outsourcing management functions is the most appropriate way to achieve best value or best performance in terms of the long-term achievement of the agency's mission.

OUTSOURCING DECISION

In determining whether a management function should be the responsibility of in-house staff or outsourced to an external organization, the agency should consider the following factors, any one of which might be key, depending on the agency and its operational circumstances. Every situation is unique and has its own combination of resources, technology, organizational structure, budget constraints, and, perhaps, physical location.

Availability and Quality of Contract Services

- Are the management services readily available in other government agencies or in the marketplace? What do their performance records show?
- Should the management function remain under the direct control of the agency to avoid giving the service provider an unfair advantage over competitors for other functions or services in the project development cycle?

- Is the management function one that, by its nature, will not allow for open, discretionary review by agency staff?

An agency should first determine if the necessary management services are available from other federal agencies or in the private sector at the location where they are required. If not, outsourcing to an external organization may not be an option. "Where there is little competition in the private sector prior to privatization, a viable bidding process may not take place such that the saving anticipated or the quality improvement desired may not take place" (Seidenstadt, 1999). The agency should determine, however, if the services are not available at all or are simply not available temporarily because of current peak activity in the economy. The agency should also determine whether the management function is a one-time project assignment or a continuing function for which it would be worth the effort of an external organization to establish a new presence based on the likelihood of subsequent contracts. The marketplace normally responds to a long-term need if other factors show that outsourcing would be appropriate.

Cost Effectiveness

- Will this management function be performed often enough within the agency that obtaining the specific expertise and associated efficiency in house would result in higher quality or be more cost effective than outsourcing the function?
- Will oversight by government staff to ensure cost effectiveness and acceptable quality be so extensive and expensive that management of the function by in-house staff would be more cost effective?

If either an external organization or in-house staff could feasibly manage the given function, cost effectiveness should be a major consideration. Cost effectiveness can be determined by a comparative cost analysis of agency costs and those of other federal agencies or private-sector firms. A "major national study suggests that the cost of in-house service delivery is frequently underestimated by as much as 30 percent. At the same time, the cost of buying services from a private vendor is often underestimated owing to a failure to account for such government costs as contract administration and contract monitoring" (Seidenstadt, 1999). The most difficult part of this analysis is ensuring that the cost proposals are truly equivalent for purposes of comparison. Typically, private-sector firms and public agencies use different accounting procedures for determining overhead costs, personnel costs, and other costs. Therefore, in making cost comparisons, overhead and other costs must somehow be displayed in equivalent terms. The agency costs of preparing the contract packages (e.g., specifications, Request for Proposals, review costs, etc.), as well as agency costs for administering the final contract, should be considered part of the contract costs.

In the committee's opinion, the savings to the agency by outsourcing the function should only be measured in "hard," actual cash-flow savings earned by moving the function outside the agency.

In calculating potential savings through staff reductions, the agency should include the costs of employee termination, which may be substantial and can result in strong negative community pressures. In-house staff represent a significant investment in time and training that should be accounted for. Agencies should carefully consider an action that results in a direct loss of experienced, trained staff in favor of relatively less experienced contract personnel. However, if existing staff can be reassigned to similar work or to fill other vacancies, or if they may actually be hired by the external organization, the impact on the agency may be minimal.

Timeliness

- Would it take more time to procure the management function from an external organization than for in-house staff to handle it?

It takes time to prepare and review contract specifications and proposals, analyze those proposals, interview proposers, and negotiate a contract with an external organization. In some cases, urgency may necessitate that the management function be assigned to in-house staff. If in-house staff has limited time, the agency should consider having in-house staff handle urgent projects and outsourcing the management of less time-sensitive projects. If an external organization has fully qualified personnel available and established procedures and/or equipment and can offer prompt service, outsourcing may be an appropriate decision.

Risk Management

- Is the management function so critical to the project that the project will fail if a failure goes undetected for an extended period of time?
- If the management function is outsourced on a constant, long-term basis, will a skill vacuum be created in the agency that would jeopardize its ability to conduct future oversight responsibilities?
- Will the performance of the management function by an external organization increase the overall costs because of liability concerns or third-party fiduciary requirements?

Management functions, by definition, control project accomplishment. For a federal agency, oversight of an external organization managing a project is less direct than oversight of the project by in-house staff. Some projects may be so critical to the performance of an agency's mission that agency responsibility and accountability virtually mandate in-house management. Facilities built in support of ongoing defense initiatives or that involve critical security or political risks may be in this category. A decision to retain a management function in house for reasons of mission-criticality should be accompanied by a clear statement of the critical aspects of the project that require in-house management control and the reasons an external organization could not reasonably meet those critical requirements.

Risk is defined as exposure to the possibility of injury or loss. Special considerations may preclude the outsourcing of certain functions (e.g., those that involve sensitive national security operations, facilities, or products or the retention of corporate memory of experienced personnel). For example, recovery from the bomb blast in the World Trade Center was greatly accelerated by the resident knowledge of the agency facilities staff for that building (Marchese, 1993). In considering outsourcing management functions, an agency should first identify special considerations and the reasons some functions should be managed by in-house staff.

The objective of risk management is to minimize the probability or magnitude of undesired consequences without incurring excessive costs (Moavenzadeh, 1997). The potential liability and risk management issues of any contract function should be examined closely. The external organization is an additional agent involved in an agency's overall performance and will not be as directly controlled as the agency's in-house staff, thus increasing potential liability exposure. The proper bundling of risk-management features in a turnkey project, however, may reduce risk for an agency. The risk-management aspect of outsourcing should be included in any contract arrangement.

RESPONSIBILITIES-AND-DELIVERABLES MATRIX

Once a decision has been made to outsource some or all of the management functions for a facility acquisition, an agency should clearly define the roles and responsibilities of all of the entities involved. The committee recommends that federal agencies establish a responsibilities-and-deliverables[1] matrix to help eliminate overlapping responsibilities, provide accountability, and ensure that, as problems arise, solutions are effectively managed. The matrix will vary from project to project, depending on the management functions outsourced and the type of contract used. Figure 4-2 is an example of this type of matrix.

---

[1]Deliverables are any measurable, tangible, verifiable outcome, result, or item that must be produced to complete a project or part of a project (PMI, 1996).

RESPONSIBILITIES-AND-DELIVERABLES MATRIX	User Management	Owner Management	Owner Project Manager	Outsourced Project Management	(A – E)	Construction Contractor	Specialty Contractors
Programming Phase							
Project request	A	P	S	S			
Deliverables/responsibilities package			A	P			
Conceptual Planning Phase							
Architect-engineer contracts			A		P		
Detailed requirements	R	A	P	S	S		
Design Phase							
Conceptual and schematic designs	R	C	A	C	P		
Permits			A	C	P		
Design development	A	A	A	C	P		
Construction documents			C	A	P		S
Procurement Phase							
List of bidders and requests for proposals			A	C	P		
Proposals (submitted)			A	C	S	P	P
Contract for construction			A	P	S	S	S
Construction Phase							
Construction permits			A	C		P	
Construction management			C	A	S	P	P
Construction work			C	A	S	P	P
Final payment (construction complete)		A	A	C	S	P	P
Start-up Phase							
Equipment installation				C	A	S	P
Move administration		P	S	S			S
Final acceptance	A	A	C	P		S	
Closeout Phase			P	S	S	S	S

FIGURE 4-2 Example of a responsibilities-and-deliverables matrix.

Note: P = primary responsibility
A = approve (signing of approval)
C = concurrence
R = reviews (no response required)
S = support (uses own resources)

PERFORMANCE MEASURES FOR EVALUATING OUTSOURCING DECISIONS

One component of the committee's statement of task was to identify measures to determine performance outcomes for outsourced management functions for facility acquisition programs (as opposed to projects). A key element of organizational decision making is measuring the effectiveness of those decisions, both qualitatively and quantitatively. The Government Performance and Results Act of 1993, which applies to all executive agencies, requires that they develop measures to determine the effectiveness of their programs and activities. These measures are normally derived from the goals and objectives of the agency's mission.

According to Independent Project Analysis, Inc., a firm of international project management specialists, effective performance measures should be: (1) related to bottom-line performance (however defined by the owner); (2) measurable and readily available; and (3) serve a clear purpose (Hess, 1997). A single measure rarely, if ever, tells the whole story of a program because individual project performance involves not only schedule or cost but also management factors and how they interrelate. One of the owner's responsibilities is to define measures that provide a detailed enough picture of program performance to enable meaningful intervention on the part of the owner or manager. Ideally, an analysis of the performance measures will enable the owner to identify which aspects of the outsourcing strategy are working well and which are not.

When management functions for facility acquisitions are outsourced, the principal measures of effectiveness of the entire program and of individual projects should relate to cost, schedule, and safety of the projects, as well as to the functionality and overall quality of the acquired facilities. Figure 4-3 is an example of a simple set of performance measures for individual projects comparing actual costs to estimated costs, actual schedules to estimated schedules, and absolute costs to the costs of similar projects.

In an earlier NRC study of performance measures for infrastructure systems, measures for these systems were grouped into categories of reliability, effectiveness, and cost (NRC, 1995). A system that reliably delivers an acceptable level of desired services at reasonable cost would be judged to be performing well. A similar framework using appropriate measures could be used for evaluating programs with outsourced management functions. For example, stated objectives of an outsourcing program could be to maintain existing project delivery schedules or not to exceed fixed-cost baselines by more than 5 per cent. These types of measures could assist an agency in analyzing the performance of projects for office buildings separately from projects for housing, industrial, or high-tech research facilities. The performance measures would be based on project delivery times and fixed costs, respectively.

However, measures for any performance-based objectives could be developed. These could include measures related to relationships between the agency

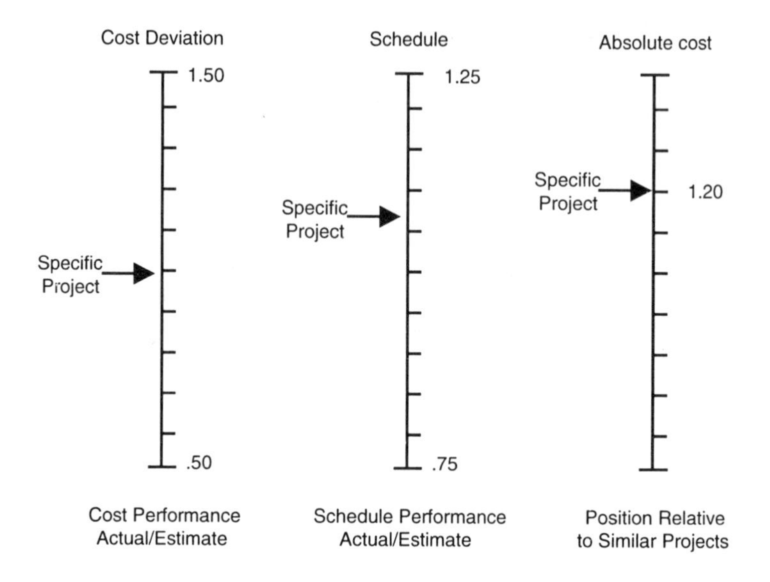

FIGURE 4-3 Simple measures of project performance. Source: Hess, 1997.

and its external organizations, for example, measures of knowledge transfer, or measures of personnel turnover. Once measures have been developed, agencies should regularly monitor and evaluate their outsourcing efforts to determine if they are meeting established objectives and identify the factors that are key to their success or failure.

Baselines and Benchmarks

If baseline levels of service have already been developed or can be developed empirically, comparing the measures and determining how well outsourcing meets the basic level of expectation should be straightforward. The committee recognizes that some federal agencies may not have baseline data on current or past performance relevant to facility programs. If no baseline exists, one should be developed for effective performance measurement.

Figure 4-4 shows how performance measures can be used to evaluate the performance of a project in comparison to the performance of the agency's entire program. If projects are categorized by their use of in-house staff or external organizations, relatively simple comparisons could indicate how the outsourced projects are performing in comparison to a baseline of government-provided services.

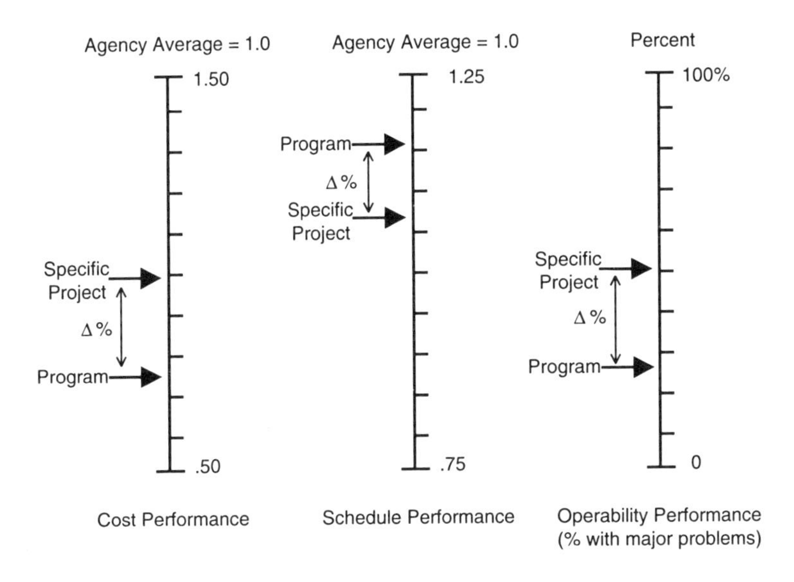

FIGURE 4-4 Project performance measured against agency baselines (for illustration purposes only).

Individual performance measures should be developed by the agencies that will use them and should not be prescribed by higher levels of government. Although it is entirely appropriate that operational guidance requiring the use of performance measures be promulgated government-wide (e.g., Government Performance and Results Act), and that the characteristics to be measured be addressed, the parties responsible for the provision of a service are in the best position to determine what constitutes good performance (NRC, 1995). An agency that decides to outsource management functions for planning, design, and construction services should be prepared to develop and apply meaningful performance measures to determine if the agency is meeting its stewardship responsibilities.

INTERAGENCY COOPERATION

Federal agencies involved in facilities acquisition, whether as owners, users, and/or providers, operate in a dynamic environment. Missions are being reevaluated, business processes are being reengineered, staffs are being downsized, and procurement processes are being modified as part of government-wide efforts to improve critical areas of performance, such as quality, cost, delivery time, and customer service. Interagency cooperation and coordination can contribute to these efforts by identifying processes and procedures that have been shown to be efficient and cost effective and implementing them. Sharing lessons learned through networking, either face-to-face or electronically (e.g., via the Internet), can be an effective method of identifying best practices for facility acquisition and for outsourcing management functions. By sharing experiences, agencies can adapt successful practices to their situation and avoid practices that have been unsuccessful. If outsourcing of management functions by federal agencies becomes commonplace, consideration should be given to the creation of a government-wide database of performance information.

SUMMARY

Chapter 4 provides a decision framework for federal agencies considering outsourcing management functions for facility acquisitions. Based on the constraints of inherently governmental functions, the framework incorporates the committee's two-step threshold test for identifying ownership functions, which should be performed by in-house staff, and management functions, which can be considered for outsourcing. An agency should first determine whether a function is inherently governmental or one that can legally be outsourced. The next step is to determine if the function is an ownership function that involves decision making on important issues or a management function that involves ministerial tasks. If it is an ownership function, it should not be outsourced.

For management functions, the next step in the decision process is to determine whether there is a need for or advantage to outsourcing. The key factors to be considered are the agency's role as owner, user, or provider of facilities, its core competencies, and the availability of in-house staff to perform the function. The last step in the decision process is to determine if outsourcing is an appropriate way to proceed. Factors in this decision will include the availability and quality of contract services, cost effectiveness, timeliness, and risk.

Once a decision has been made to outsource some or all of the management functions for facility acquisition, an agency should clearly define the roles and responsibilities of all entities involved. The committee recommends that federal agencies establish a responsibilities-and-deliverables matrix to help eliminate overlapping responsibilities, provide accountability, and ensure that solutions are managed effectively as problems arise.

A key element of organizational decision making is measuring the effectiveness of those decisions, both qualitatively and quantitatively. When management functions for facility acquisition are outsourced, the principal measures of effectiveness of the entire program and of individual projects should relate to cost, schedule, and safety of the projects, as well as the functionality and overall quality of the acquired facilities.

Individual performance measures should be developed by the agencies that will use them. An agency that decides to outsource management functions for planning, design, and construction services should be prepared to develop and apply meaningful, quantifiable performance metrics to determine if the agency is meeting its stewardship responsibilities.

Federal agencies involved in facility acquisitions operate in a dynamic environment. Sharing lessons learned through networking can be an effective method of identifying best practices for facility acquisition and for outsourcing management functions.

FINDINGS AND RECOMMENDATIONS

Finding. The outsourcing of management functions for planning, design, and construction-related services by federal agencies is a strategic decision that should be considered in the context of an agency's long-term mission.

Finding. Performance measures are necessary to assess the success of any outsourcing effort.

Recommendation. A federal agency should analyze the relationship of outsourcing decisions to the accomplishment of its mission before outsourcing management functions for planning, design, or construction services. Outsourcing for services and functions should be integrated into an overall strategy for achieving the agency's mission, managing resources, and obtaining best value or best performance for the resources expended. Outsourcing of management functions should not be used solely as a short-term expedient to limit spending or reduce the number of in-house personnel.

Recommendation. Once a decision has been made to outsource some or all management functions, a responsibilities-and-deliverables matrix should be established to help eliminate overlapping responsibilities, provide accountability and ensure that, as problems arise, solutions are managed effectively.

Recommendation. Agencies that outsource management functions for planning, design, and construction services should regularly evaluate the effectiveness of the outsourcing effort in relation to accomplishment of the agency's mission.

Recommendation. Agencies should establish performance measures to assess accomplishments relative to the objectives established for the outsourcing effort and, at a minimum, address cost, schedule, and quality parameters.

Recommendation. Interagency coordination, cooperation, collaboration, networking, and training should be increased to encourage the use of best practices and improve life-cycle cost effectiveness in the delivery of federal facilities.

REFERENCES

Hess, R. 1997. Establishing Performance Measures for Outsourcing Projects. Presentation by Ronald Hess, Independent Project Analysis, to the Committee on Outsourcing of Design and Construction Management-Related Activities for Federal Facilities, National Research Council, Washington, D.C., September 22, 1997.

Marchese, M. 1993. World Trade Center Disaster. EPICCGRAM Fall 1993 (#3). New York: Emergency Preparedness for Industry and Commerce Council.

Moavenzadeh, F. 1997. Risk management. Construction Business Review 7(4): 5.

NRC (National Research Council). 1995. Measuring and Improving Infrastructure Performance. Board on Infrastructure and the Constructed Environment, National Research Council. Washington, D.C.: National Academy Press.

PMI (Project Management Institute). 1996. A Guide to the Project Management Body of Knowledge. Sylva, N.C.: PMI Communications.

Seidenstadt, P., ed. 1999. Contracting Out Government Services: Theory and Practice in the United States. Westport, Conn.: Praeger.

Appendixes

Appendix A

Biographical Sketches of Committee Members

Henry L. Michel (chair) was elected to the National Academy of Engineering in 1995 for leadership in applied-research technology transfer and promoting alternative forms of project execution. Mr. Michel is Chairman Emeritus of Parsons, Brinckerhoff, Inc. His professional career encompasses 50 years of highly diversified engineering experience focused mostly on transportation planning, rail and rapid-transit system design, and construction management. He has served as principal-in-charge for major, urban, rapid-transportation projects, including the Metropolitan Atlanta Rapid Transit Authority (MARTA) system, the Caracas Metro, and the development of a new rapid-transit system for Taipei, Taiwan.

Mr. Michel is the author of numerous technical articles and an active member of many professional engineering societies at the state and national levels. He has served as chair of the International Road Federation, chair of the Civil Engineering Research Foundation, and vice chair of the Building Futures Council and has received many professional awards. He is a fellow of the American Society of Civil Engineers, the American Consulting Engineers Council, the Institute of Civil Engineers, and the Society of Military Engineers. Mr. Michel has a B.S. in civil engineering from Columbia University and is a registered professional engineer in New York and Pennsylvania.

Joseph A. Ahearn is president of the Transportation Business Group, CH2M Hill, a global infrastructure and engineering firm. Major General Ahearn, USAF (retired), served as eastern regional manager and federal programs manager in the Environmental Business Line in previous assignments at CH2M Hill. He has also conducted strategic planning and positioning initiatives on advanced project-delivery systems for industrial and governmental clients.

Prior to his position with CH2M Hill, General Ahearn completed 34 years of military service, ending his career as the U.S. Air Force civil engineer. His military assignments included management of the construction of missile launch facilities throughout the United States; director of airfield operations and maintenance in Labrador; construction of intelligence facilities in Europe and Asia; airfield construction leadership in Vietnam; worldwide Inspector General duty; management of engineering, construction, and operations for the entire European NATO basing; and planning, design, and construction of the basing network during Desert Storm.

General Ahearn is involved in a number of professional architectural and engineering societies. He has served as vice chair of the Board of Directors of the Civil Engineering Research Foundation and chair of the Academy of Fellows, Society of Military Engineers. He was also a member of the Building Futures Council of the National Academy of Engineering. General Ahearn holds a B.S. in civil engineering from the University of Notre Dame, an M.S. in engineering administration from Syracuse University, and is a registered professional engineer in the state of Massachusetts.

A. Wayne Collins is the deputy state engineer for planning and engineering, Arizona Department of Transportation. Mr. Collins' responsibilities include oversight of all intermodal infrastructure planning, programming, scheduling, and design performance in the state of Arizona, with the exception of surface state highway programs in the Phoenix metropolitan area. Mr. Collins is a registered civil engineer in the District of Columbia and Arizona, a registered land surveyor in Arizona, and a member of the American Institute of Certified Planners. He is former president of the American Public Works Association; past president, Arizona Section, American Society of Civil Engineers (ASCE); and past president, National Association of County Engineers. His professional affiliations include Arizona Society of Professional Engineers, and National Society of Professional Engineers, Society of American Military Engineers, American Planning Association, Institute of Traffic Engineers, and the Governor's Solid Waste Management Board.

In 1996, the American Public Works Association named Mr. Collins one of the Nation's "Top Ten" Public Works Leaders. Other awards include the ASCE Zone IV (Western U.S.) Government Civil Engineer of the Year Award for 1987; Government Engineer of the Year for 1990 from the Arizona Society of Professional Engineers; and Distinguished Service Award for 1993 from the Arizona Section of ASCE.

John D. Donahue is associate professor of public policy at the John F. Kennedy School of Government, Harvard University. Dr. Donahue's research encompasses issues that arise along the boundary between the public and private sectors. His current work concerns the definition of government's proper role in promoting a nation's economic policies at all levels of government. In January 1993, Dr. Donahue was appointed assistant secretary of labor; he became counselor to the secretary of labor in 1994 and served in that position until August 1995. His book, *The Privatization Decision: Public Ends, Private Means* (Basic Books, 1989), has been translated into four languages. Dr. Donahue is coauthor, with Robert B. Reich, of *New Deals: The Chrysler Revival and the American System* (Times Books, 1985), and his articles and essays have appeared in *Atlantic Monthly, Washington Monthly,* and *The New York Times Book Review.* He holds a B.A. from Indiana University and an M.A. and Ph.D. in public policy from Harvard University.

Lloyd A. Duscha was elected to the National Academy of Engineering in 1987 for distinguished engineering and construction administration of water-resources projects and military facilities. Mr. Duscha retired from the U.S. Army Corps of Engineers in 1990 after serving as deputy director of engineering and construction. He is currently an engineering consultant to government agencies and private-sector clients. His experience encompasses policy development, organizational management, planning and programming, design and construction, project management, and contract administration. He has served on numerous National Research Council (NRC) committees, including the Committee to Assess the Policies and Practices of the Department of Energy in Project Management and the Committee on Shore Protection Readiness and Management. He is a past member of the NRC Board on Infrastructure and the Constructed Environment and was vice chair of the U.S. National Committee on Tunneling Technology. He holds a B.C.E. from the University of Minnesota, which also awarded him the Board of Regents Outstanding Achievement Award.

G. Brian Estes is the former director of construction projects, Westinghouse Hanford Company, where he directed project management for construction projects in support of operations and environmental cleanup of the U.S. Department of Energy Hanford Nuclear Complex. Prior to joining Westinghouse, Mr. Estes served for 30 years in the Navy Civil Engineer Corps, achieving the rank of rear admiral. Admiral Estes served as commander of the Pacific Division of the Naval Facilities Engineering Command and commander of the Third Naval Construction Brigade, Pearl Harbor. He supervised more than 700 engineers, 8,000 Seabees, and 4,000 other employees in providing public works management, environmental support, family housing support, facility planning, and design and construction services. As vice commander, Naval Facilities Engineering Command, Admiral Estes led the total quality-management transformation at headquarters and two updates of the corporate strategic plan. While he was commander for facilities acquisition and deputy commander for public works, Naval Facilities Engineering Command, he directed the execution of the $2 billion military construction program and the $3 billion facilities-management program. He holds a B.S. in civil engineering from the University of Maine and an M.S. in civil engineering from the University of Illinois. He is a registered professional engineer in Illinois.

Mark C. Friedlander, a partner in the construction law group of the law firm of Schiff, Hardin, and Waite, is past chair of the Professional Practices and Contracts Committee of the Design Build Institute of America. He is an adjunct professor at the University of Illinois at Chicago School of Architecture and has been a lecturer and instructor at the Georgia Institute of Technology and the Northwestern University School of Engineering, the University of Wisconsin at Madison, and the Illinois Institute of Technology. Mr. Friedlander has been a member of the Construction Arbitration Advisory Panel of the American Arbitration Association and a member of the Forum Committee on Construction Law, as well as the Chicago Bar Association and American Bar Association. He has published and presented numerous papers related to construction law, design build, professional liability, and the legal responsibilities of architects and engineers. Mr. Friedlander holds a B.A. from the University of Michigan and a J.D. from Harvard Law School.

Henry J. Hatch was elected to the National Academy of Engineering in 1992 for leadership in the engineering and construction programs of the U.S. Army Corps of Engineers and exceptional management of its programs. He was the chief operating officer of ASCE from 1997 to 1999. Before joining ASCE, he was president and chief executive officer of Fluor Daniel, Hanford, Inc., where he directed a $5 billion, five-year management contract for the U.S. Department of Energy's environmental cleanup at Hanford. Prior to joining Fluor, he was president and chief operating officer of Law Companies Group, Inc., a worldwide engineering and environmental-services company. In 1992, Mr. Hatch completed a distinguished 35-year career in the United States Army; where he achieved the rank of lieutenant general and served as chief of engineers and commander of the U.S. Army Corps of Engineers. As chief of engineers, General Hatch commanded more than 40,000 members of the Corps and supervised an annual budget of more than $13 billion. General Hatch served in a variety of important command and staff assignments, including engineer for the U.S. Army in Europe; commander of the Corps' Pacific Ocean Division in Hawaii; commander of the Corps' Nashville District; commander of the 326th Engineer Battalion, 101 Airborne Division, in Vietnam; and instructor and assistant professor at West Point.

General Hatch is a registered professional engineer in the District of Columbia and a member and past national president of the Society of American Military Engineers. He received the President's Award from ASCE in 1991, the Chairman's Award from the Natural Resources Council of America in 1992, and was recognized as one of the Nation's Top 10 Public Works Leaders in 1990 by the American Public Works Association. General Hatch graduated from the U.S.

Military Academy at West Point and has an M.S. in geodetic science from Ohio State University.

Stephen C. Mitchell is president and chief operating officer of Lester B. Knight and Associates, Inc., a privately held, professional services company. As chair of Knight's Chicago practice, Knight Architects Engineers and Planners, Inc., Mr. Mitchell is involved in all aspects of the firm's management and has been the principal strategic planning officer. He is a member of the Board of Directors of Apogee Enterprises, Inc., a world leader in the design, fabrication, and erection of building curtain-wall systems. He is also on the Board of Directors of Delon Hampton Associates, Ltd., an engineering consulting firm. Mr. Mitchell has participated in a number of science and technology activities for the state of Illinois, including the Governor's Science and Technology Task Force and its successor commissions, the Illinois Coalition, and the Governor's Science Advisory Board. He has served on the Northwestern University McCormick School of Science and Engineering Advisory Committee and Civil Engineering Advisory Committee. He is a member of Northwestern University's Trustee Associates Board. Mr. Mitchell has been continuously involved in professional societies, including ASCE and the Civil Engineering Research Foundation, and he was awarded the ASCE William Wisely American Civil Engineer Award. He holds a B.S. and M.S. in civil engineering from the University of New Mexico and an M.B.A. from the University of Chicago.

Karla Schikore has been in the corporate real estate field for more than 25 years and is currently president of KSA, Inc., a corporate real estate consulting firm. She began her career with the General Services Administration, after which she worked for Bank of America's Corporate Real Estate Group for 20 years. She has extensive background in all facets of real estate and management practices and has specialized for the past several years in global business practices related to workplace strategy, construction management, financial performance of strategic alliances, data-center management programs, and outsourcing of facilities management on a global basis. She has given more than 60 presentations and educational seminars in the last three years to real estate, design and construction, and energy-management organizations and boards. Ms. Schikore studied business administration and economics at San Francisco State University and is a certified property manager. She has also served on the Board of Directors, Board of Trustees and Research Advisory Council for the International Development Research Council, and is a member of the National Association of Corporate Real Estate Executives, Industrial Development Research Council, Urban Land Institute, Institute of Real Estate Management, and the San Francisco Board of Realtors.

E. Sarah Slaughter is the founder, chair, and president of MOCA, Inc., a provider of simulation systems to control and manage the design and construction of complex, high-performance facilities. She is currently on leave as a faculty member in the Department of Civil and Environmental Engineering at the Massachusetts Institute of Technology (MIT). Her specialty is innovation in the design and construction of built facilities, particularly system and intersystem impacts. She is a member of the NRC Board on Infrastructure and the Constructed Environment, an advisor to the Governor's Construction Reform Task Force for the Commonwealth of Massachusetts, and a team leader for the Innovation Systems in Construction Task Group for the Conseil Internationale du Batiment pour la Recherche l'Etude et la Documentation. Her current research projects include the development of dynamic process-simulation models of system and material-specific construction activities to assess innovations, analysis, the development of design strategies to accommodate changes in built facilities over the long term, and theory development and analysis of effective collaboration mechanisms between organizations for the development and use of innovations. Dr. Slaughter is the recipient of the Gilbert Winslow Career Development Chair at MIT and has received the CAREER Program Award from the National Science Foundation. She is a member of Sigma Xi, the National Society of Professional Engineers, ASCE, and the American Society of Engineering Education. She served on the NRC Committee for an Infrastructure Technology Research Agenda. Dr. Slaughter earned an S.B. in civil engineering and anthropology, an S.M. in civil engineering and technology policy, and a multidisciplinary Ph.D. in the management of technology from MIT.

Luis M. Tormenta is the chairman and chief operating officer with The LIRO Group. Previously, he was vice president and general manager of Raytheon Infrastructure, Inc. Mr. Tormenta was formerly commissioner of the New York City Department of Design and Construction through 1999. He was appointed to that post by Mayor Giuliani to create and manage a "super construction agency" to serve as the primary vehicle in the delivery of the capital construction program. Mr. Tormenta was responsible for an operating budget of more than $72 million, a capital budget of $3 billion, and a staff of more than 1,400. He was directly responsible for the development of the department's organizational structure and for the establishment of the agency's operating methodology and philosophy. Under his guidance, independent operations of various mayoral agencies were consolidated, and numerous bureaucratic processes were streamlined and reengineered. As director of Design and Construction, Facilities Development Corporation, Mr. Tormenta reduced project durations by 30 percent through the establishment of management teams and implemented an outreach program to deal with local and minority issues. He earned a B.S. in civil engineering from Manhattan College and is a licensed professional engineer in the state of New York.

Richard L. Tucker was elected to the National Academy of Engineering in 1996 for developing management-improvement practices in construction. Dr. Tucker is the director of the Center for Construction Industry Studies, holds the Joe C. Walter Chair in engineering at the University of Texas at Austin, and is past director of the Construction Industry Institute. His primary research interest is in project management, including all aspects of capital facilities delivery, from conception to successful operation. His research focuses on the development of effective tools for preproject planning, improving efficiency and effectiveness of the design and procurement processes, and improving construction productivity, as well as the development of methods and metrics for benchmarking and measuring the facility-delivery process. Dr. Tucker has been a member of two NRC committees, the Committee for the Study of Approaches for Increasing Private-Sector Involvement in the Highway Innovation Process and the Committee on Construction Productivity. Dr. Tucker served as a member of the Board of Directors for Hill and Wilkinson, Inc., Integrated Electrical Services, and Tucker and Tucker Consultants. He is a fellow of the ASCE and a member of numerous professional societies and associations, including the National Society of Professional Engineers, American Association of Cost Engineers, and the American Society for Testing and Materials. Dr. Tucker's awards and honors include the Construction Engineering Educator Award for Individual Initiative, Construction Industry Institute, 1994; and the Michael Scott Endowed Research Fellow, Institute for Constructive Capitalism, 1990. His numerous publications span four decades. Dr. Tucker has a B.S., M.S., and Ph.D. in civil engineering from the University of Texas at Austin.

Norbert W. Young, Jr., is president of the McGraw-Hill Construction Information Group, a leading source of project news, product information, industry analysis, and editorial coverage for design and construction professionals. Mr. Young joined the McGraw-Hill Companies in November 1997 as vice president, editorial, for F.W. Dodge. Before joining McGraw-Hill, he spent eight years with the Bovis Construction Group, a global leader in the management of high-profile construction projects. In 1994, he was appointed president of Bovis Management Systems. Notable clients and projects with which Mr. Young was involved include the 1996 Summer Olympic Games, Bank of America, NYNEX, and Sun Microsystems. Mr. Young has more than 25 years of design and construction experience and is a registered architect in Pennsylvania, Connecticut, Maryland, and Maine. During the 1980s, Mr. Young was a partner at Toombs Development Company, a leading real estate firm in New Canaan, Connecticut. He managed all aspects of design and construction, including governmental permitting and approvals for the firm's real estate developments in Philadelphia, Baltimore, and New Jersey. He holds an M. Arch. from the University of Pennsylvania and a B.A. from Bowdoin College. His professional affiliations include membership in the Urban Land Institute, American Institute of Architects, the International Alliance for Interoperability, and the International Development Research Council.

Appendix B

List of Briefings

National Research Council, Washington, D.C., June 2, 1997

Briefings by senior representatives of sponsoring agencies:

Peter Devlin
Project Manager, Office of Field Management
U.S. Department of Energy

Carl A. Petchik
Branch Chief, Africa and Eastern Pacific Projects
Office of Foreign Building Operations
U.S. Department of State

Robert L. Neary, Jr.
Deputy Facilities Management Officer, Office of Facilities Management
U.S. Department of Veterans Affairs

National Research Council, Washington, D.C., September 22–23, 1997

Panel Session: Outsourcing Initiatives: Federal and Private-Sector Experiences

Presentations:

Colonel Joseph Munter
Director of Outsourcing and Privatization, Office of the Civil Engineer
U.S. Air Force

Walter Cheatham
Engineering Division
DuPont Company, Inc.

Tom Graves
Director, Program Development Division, Public Buildings Service
General Services Administration

Charles Kluenker
President, Western Region
3D International, Inc.

Ronald Hess
Project Manager
Independent Project Analysis, Inc.

David Childs
A-76 Program Examiner
Office of Management and Budget
Executive Office of the President

Bill Keating
KPMG Peat Marwick, Inc.

M. Rolle Walker
AON Construction Services Division
AON Risk Services, Inc.

Appendix C

Documents Related to Inherently Governmental Functions

September 23, 1992

POLICY LETTER 92-1
TO THE HEADS OF EXECUTIVE AGENCIES AND DEPARTMENTS

SUBJECT: Inherently Governmental Functions

1. **Purpose**. This policy letter establishes Executive Branch policy relating to service contracting and inherently governmental functions. Its purpose is to assist Executive Branch officers and employees in avoiding an unacceptable transfer of official responsibility to Government contractors.

2. **Authority**. This policy letter is issued pursuant to section 6(a) of the Office of Federal Procurement Policy (OFPP) Act, as amended, codified at 41 U.S.C. [[section]] 405.

3. **Exclusions**. Services obtained by personnel appointments and advisory committees are not covered by this policy letter.

4. **Background**. Contractors, when properly used, provide a wide variety of useful services that play an important part in helping agencies to accomplish their missions. Agencies use service contracts to acquire special knowledge and skills not available in the Government, obtain cost effective services, or obtain temporary or intermittent services, among other reasons.

   Not all functions may be performed by contractors, however. Just as it is clear that certain functions, such as the command of combat troops, may not be contracted, it is also clear that other functions, such as building maintenance and food services, may be contracted. The difficulty is in determining which of these services that fall between these extremes may be acquired by contract. Agencies have occasionally relied on contractors to perform certain functions in such a way as to raise questions about whether Government policy is being created by private persons. Also, from time to time questions have arisen regarding the extent to which de facto control over contract performance has been transferred to contractors. This policy letter provides an illustrative list of functions, that are, as a matter of policy, inherently governmental (see Appendix A)(, and articulates the practical and policy considerations that underlie such determinations (see [[section]] 7).

   As stated in [[section]] 9, however, this policy letter does not purport to specify which functions are, as a legal matter, inherently governmental, or to define the factors used in making such legal determination. Thus, the fact that a function is listed in Appendix A, or a factor is set forth in [[section]] 7(b), does not necessarily mean that the function is inherently governmental as a legal matter or that the factor would be relevant in making the legal determination.

5. **Definition**. As a matter of policy, an "inherently governmental function" is a function that is so intimately related to the public interest as to mandate performance by Government employees. These functions include those activities that require either the exercise of discretion in applying Government authority or the making of value judgments in making decisions for the Government. Governmental functions normally fall into two categories: (1) the act of governing, i.e., the discretionary exercise of Government authority, and (2) monetary transactions and entitlements.

   An inherently governmental function involves, among other things, the interpretation and execution of the laws of the United States so as to:

(a) bind the United States to take or not to take some action by contract, policy, regulation, authorization, order, or otherwise;

(b) determine, protect, and advance its economic, political, territorial, property, or other interests by military or diplomatic action, civil or criminal judicial proceedings, contract management, or otherwise;

(c) significantly affect the life, liberty, or property of private persons;

(d) commission, appoint, direct, or control officers of employees of the United States; or

(e) exert ultimate control over the acquisition, use, or disposition of the property, real or personal, tangible or intangible, of the United States, including the collection, control, or disbursement of appropriated and other Federal funds.

Inherently governmental functions do not normally include gathering information for or providing advice, opinions, recommendations, or ideas to Government officials. They also do not include functions that are primarily ministerial and internal in nature, such as building security; mail operations; operation of cafeterias; housekeeping; facilities operations and maintenance, warehouse operations, motor vehicle fleet management and operations, or other routine electrical or mechanical services.

The detailed list of examples of commercial activities found as an attachment to Office of Management and Budget (OMB) Cir. No. A-76 is an authoritative, nonexclusive list of functions that are not inherently governmental functions. These functions therefore may be contracted.

6. **Policy.**

(a) **Accountability**. It is the policy of the Executive Branch to ensure that Government action is taken as a result of informed, independent judgments made by Government officials who are ultimate accountable to the President. When the Government uses service contracts, such informed, independent judgment is ensured by:

(1) prohibiting the use of service contracts for the performance of inherently governmental functions (See Appendix A);

(2) providing greater scrutiny and an appropriate enhanced degree of management oversight (see subsection 7(f)) when contracting for functions that are not inherently governmental but closely support the performance of inherently governmental functions (see Appendix B);

(3) ensuring, in using the products of those contracts, that any final agency action complies with the laws and policies of the United States and reflects the independent conclusions of agency officials and not those of contractors who may have interests that are not in concert with the public interest, and who may be beyond the reach of management controls otherwise applicable to public employees; and

(4) ensuring that reasonable identification of contractors and contractor work products is made whenever there is a risk that the public, Congress, or other persons outside of the Government might confuse them with Government officials or with Government work products, respectively.

(b) **OMB Circular No. A-76**. This policy letter does not purport to supersede or otherwise effect any change in OMB Circular No. A-76, Performance of Commercial Activities.

(c) **Drafting of Congressional testimony, responses to Congressional correspondence, and agency responses to audit reports from an Inspector General, the General Accounting Office, or other Federal audit entity**. While the approval of a Government document is an inherently governmental function, its drafting is not necessarily such a function. Accordingly, in most situations the drafting of a document, or portions thereof, may be contracted, and the agency should review and revise the draft document, to the extent necessary, to ensure that the final document expresses the agency's views and advances the public interest. However, even though the drafting function is not necessarily an inherently government function, it may be inappropriate, for various reasons, for a private party to draft a document in particular circumstances. Because of the appearance of private influence with respect to documents that are prepared for Congress or for law enforcement or oversight agencies and that may be particularly sensitive, contractors are not to be used for the drafting of Congressional testimony; responses to Congressional correspondence; or agency responses to audit reports from an Inspector General, the General Accounting Office, or other Federal audit entity.

7. **Guidelines**. If a function proposed for contract performance is not found in Appendix A, the following guidelines will assist agencies in understanding the application of this policy letter, determining whether the function is, as a matter of policy, inherently governmental and forestalling potential problems.

(a) **The exercise of discretion**. While inherently governmental functions necessarily involve the exercise of substantial discretion, not every exercise of discretion is evidence that such a function is involved. Rather, the use of discretion must have the effect of committing the Federal Government to a course of action when two or more alternative courses of action exist (e.g., purchasing a minicomputer than a mainframe computer, hiring a statistician rather than an economist, supporting proposed legislation rather than opposing economist, supporting proposed legislation rather than opposing it, devoting more resources to prosecuting one type of criminal case than another, awarding a contract to one firm rather than another, adopting one policy rather than another, and so forth).

A contract may thus properly be awarded where the contractor does not have the authority to decide on the course of action to be pursued but is rather tasked to develop options to inform an agency decision maker, or to develop or expand decisions already made by Federal officials. Moreover, the mere fact that decisions are made by the contractors in performing his or her duties (e.g., how to allocate the contractor's own or subcontract resources, what techniques and procedures to employ, whether and whom to consult, what research alternatives to explore given the scope of the contract, what conclusions to emphasize, how frequently to test) is not determinative of whether he or she is performing an inherently government function.

(b) **Totality of the circumstances**. Determining whether a function is an inherently governmental function often is difficult and depends upon an analysis of the factors of the case. Such analysis involves consideration of a number of factors, and the presence or absence of any one is not in itself determinative of the issue. Nor will the same emphasis necessarily be placed on any one factor at different times, due to the changing nature of the Government's requirements.

The following factors should be considered when deciding whether award of a contract might effect, or the performance of a contract has effected, a transfer of official responsibility:

(1) Congressional legislative restrictions or authorizations.

(2) The degree to which official discretion is or would be limited, i.e., whether the contractor's involvement in agency functions is or would be so extensive or his or her work product is so far advanced toward completion that the agency's ability to develop and consider options other than those provided by the contractor is restricted.

(3) In claims adjudication and related services,

(i) the finality of any contractor's action affecting individual claimants or applicants, and whether or not review of the contractor's action is **de novo** (i.e., to be effected without the appellate body's being bound by prior legal rulings or factual determinations) on appeal of his or her decision to an agency official;

(ii) the degree to which contractor activities may involve wide-ranging interpretations of complex, ambiguous case law and other legal authorities, as opposed to being circumscribed by detailed laws, regulations, and procedures.

(iii) the degree to which matters for decision by the contractor involve recurring fact patterns or unique fact patterns; and

(iv) The contractor's discretion to determine an appropriate award or penalty.

(4) The contractor's ability to take action that will significantly and directly affect the life, liberty, or property of individual members of the public, including the likelihood of the contractor's need to resort to force in support of a police or judicial function; whether force, especially deadly force, is more likely to be initiated by the contractor or by some other person; and the degree to which force may have to be exercised in public or relatively uncontrolled areas. (Note that contracting for guard, convoy security, and plant protection services, armed or unarmed, is not proscribed by these policies.)

(5) The availability of special agency authorities and the appropriateness of their application to the situation at hand, such as the power to deputize private persons.

(6) Whether the function in question is already being performed by private persons, and the circumstances under which it is being performed by them.

(c) **Finality of agency determinations**. Whether or not a function is an inherently governmental function, for purposes of this policy letter, is a matter for agency determination. However, agency decisions that a function is or is not an inherently governmental function may be reviewed, and, if necessary, modified by appropriate OMB officials.

(d) **Preaward responsibilities**. Whether a function being considered for performance by contract is an inherently governmental function is an issue to be addressed prior to issuance of the solicitation.

(e) **Post-award responsibilities**. After award, even when a contract does not involve performance of an inherently governmental function, agencies must take steps to protect the public interest by playing an active, informed role in contract administration. This ensures that contractors comply with the terms of the contract and that Government policies, rather than private ones, are implemented. Such participation should be appropriate to the nature of the contract, and should leave no doubt that the contract is under the control of Government officials. This does not relieve contractors of their performance responsibilities under the contract. Nor does this responsibility to administer the contract require Government officials to exercise such control over contractor activities to convert the contract, or portion thereof, to a personal service contract.

In deciding whether Government officials have lost or might lose control of the administration of a contract, the following are relevant considerations: the degree to which agencies have effective management procedures and policies that enable meaningful oversight of contractor performance, the resources available for such oversight, the actual practice of the agency regarding oversight, the duration of the contract, and the complexity of the tasks to be performed.

(f) **Management controls.** When functions described in Appendix B are involved, additional management attention to the terms of the contract and the manner of performance is necessary. How close the scrutiny or how extensive or stringent the management controls need to be is for agencies to determine. Examples of additional control measures that might be employed are:

(1) developing carefully crafted statements of work and quality assurance plans, as described in OFPP Policy Letter 91-2 **Service Contracting**, that focus on the issue of Government oversight and measurement of contractor performance;

(2) establishing audit plans for periodic review of contracts by Government auditors;

(3) conducting preaward conflict of interest reviews to ensure contract performance in accordance with objective standards and contract specifications;

(4) physically separating contractor personnel from Government personnel at the worksite; and

(5) requiring contractors to (a) submit reports that contain recommendations and that explain and rank policy or action alternatives, if any, (b) describe what procedures they used to arrive at their recommendations, (c) summarize the substance of their deliberations, (d) report any dissenting views, (e) list sources relied upon, and/or (f) otherwise make clear the methods and considerations upon which their recommendations are based.

(g) **Identification of contractor personnel and acknowledgment of contractor participation.** Contractor personnel attending meetings, answering Government telephones, and working in other situations where their contractor status is not obvious to third parties must be required to identify themselves as such to avoid creating an impression in the minds of members of the public or the Congress that they are Government officials, unless, in the judgment of the agency, no harm can come from failing to identify themselves. All documents or reports produced by contractors are to be suitably marked as contractor products.

(h) **Degree of reliance** The extent of reliance on service contractors is not by itself a cause for concern. Agencies must, however, have a sufficient number of trained and experienced staff to manage Government programs properly. The greater the degree of reliance on contractors the greater the need for oversight by agencies. What number of Government officials is needed to oversee a particular contract is a management decision to be made after analysis of a number of factors. These include, among others, the scope of the activity in question; the technical complexity of the project or its components; the technical capability, numbers, and workload of Federal oversight officials; the inspection techniques available; and the importance of the activity. Current contract administration resources shall not be determinative. The most efficient and cost effective approach shall be utilized.

(i) **Exercise of approving or signature authority.** Official responsibility to approve the work of contractors is a power reserved to Government officials. It should be exercised with a thorough knowledge and understanding of the contents of documents submitted by contractors and a recognition of the need to apply independent judgment in the use of these work products.

8. **Responsibilities.**

(a) **Heads of agencies.** Heads of departments and agencies are responsible for implementing this policy letter. While these policies must be implemented in the Federal Acquisition Regulation (FAR), it is expected that agencies will take all appropriate actions in the interim to develop implementation strategies and initiate staff training to ensure effective implementation of these policies.

(b) **Federal Acquisition Regulatory Council.** Pursuant to subsections 6(a) and 25(f) of the OFPP Act, as amended, 41 U.S.C. [[section]][[section]] 405(a) and 421(f), the Federal Acquisition Regulatory Council shall ensure that the policies established herein are incorporated in the FAR within 210 days from the date this policy letter is published in the **Federal Register.** Issuance of final regulations within this 210-day period shall be considered issuance "in a timely manner" as prescribed in 41 U.S.C. [[section]] 405(b).

(c) **Contracting officers.** When requirements are developed, when solicitations are drafted, and when contracts are being performed, contracting officers are to ensure:

(1) that functions to be contracted are not among those listed in Appendix A of this letter and do not closely resemble any functions listed here;

(2) that functions to be contracted that are not listed in Appendix A, and that do not closely resemble them, are not inherently governmental functions according to the totality of the circumstances test in subsection 7(b), above;

(3) that the terms and the manner of performance of any contract involving functions listed in Appendix B of this letter are subject to adequate scrutiny and oversight in accordance with subsection 7(f), above; and

(4) that all other contractible functions are properly managed in accordance with subsection 7(e), above.

(d) **All officials.** When they are aware that contractor advice, opinions, recommendations, ideas, reports, analyses, and other work products are to be considered in the course of their official duties, all Federal Government officials are to ensure that they exercise independent judgment and critically examine these products.

9. **Judicial review.** This policy letter is not intended to provide a constitutional or statutory interpretation of any kind and it is not intended, and should not be construed, to create any right or benefit, substantive or procedural, enforceable at law by a party against the United States, its agencies, its officers, or any person. It is intended only to provide policy guidance to agencies in the exercise of their discretion concerning Federal contracting. Thus, this policy letter is not intended, and should not be construed, to create any substantive or procedural basis on which to challenge any

agency action or inaction on the ground that such action or inaction was not in accordance with this policy letter.

10. **Information contact.** For information regarding this policy letter contact Richard A. Ong, Deputy Associate Administrator, the Office of Federal Procurement Policy, 725 17th Street, N.W., Washington, DC 20503. Telephone (202) 395-7209.

11. **Effective date.** This policy letter is effective 30 days after the date of publication.

(signed by)
Allan V. Burman
Administrator

APPENDIX A

The following is an illustrative list of functions considered to be inherently governmental functions: (footnote: With respect to the actual drafting of Congressional testimony, of responses to Congressional correspondence, and of agency responses to audit reports from the Inspector General, the General Accounting Office, or other Federal audit entity, see special provisions in subsection 6(c) of the text of the policy letter)

1. The direct conduct of criminal investigation.

2. The control of prosecutions and performance of adjudicatory functions (other than those relating to arbitration or other methods of alternative dispute resolution).

3. The command of military forces, especially the leadership of military personnel who are members of the combat, combat support or combat service support role.

4. The conduct of foreign relations and the determination of foreign policy.

5. The determination of agency policy, such as determining the content and application of regulations, among other things.

6. The determination of Federal program priorities or budget requests.

7. The direction and control of Federal employees.

8. The direction and control of intelligence and counter-intelligence operations.

9. The selection or nonselection of individuals for Federal Government employment.

10. The approval of position descriptions and performance standards for Federal employees.

11. The determination of what Government property is to be disposed of and on what terms (although an agency may give contractors authority to dispose of property at prices with specified ranges and subject to other reasonable conditions deemed appropriate by the agency).

12. In Federal procurement activities with respect to prime contracts,

(a) determining what supplies or services are to be acquired by the Government (although an agency may give contractors authority to acquire supplies at prices within specified ranges and subject to other reasonable conditions deemed appropriate by the agency);

(b) participating as a voting member on any source selection boards;

(c) approval of any contractual documents, to include documents defining requirements, incentive plans, and evaluation criteria;

(d) awarding contracts;

(e) administering contracts (including ordering changes in contract performance or contract quantities, taking action based on evaluations of contractor performance, and accepting or rejecting contractor products or services);

(f) terminating contracts; and

(g) determining whether contract costs are reasonable, allocable, and allowable.

13. The approval of agency responses to Freedom of Information Act requests (other than routine responses that, because of statute, regulation, or agency policy, do not require the exercise of judgment in determining whether documents are to be released or withheld), and the approval of agency responses to the administrative appeals of denials of Freedom of Information Act requests.

14. The conduct of administrative hearings to determine the eligibility of any person for a security clearance, or involving actions that affect matters of personal reputation or eligibility to participate in Government programs.

15. The approval of Federal licensing actions and inspections.

16. The determination of budget policy, guidance, and strategy.

17. The collection, control, and disbursement of fees, royalties, duties, fines, taxes and other public funds, unless authorized by statute, such as title 31 U.S.C. [[section]] 952 (relating to private collection contractors) and title 31 U.S.C. [[section]] 3718 (relating to private attorney collection services), but not including:

    (a) collection of fees, fines, penalties, costs or other charges from visitors to or patrons of mess halls, post or base exchange concessions, national parks, and similar entities or activities, or from other persons, where the amount to be collected is easily calculated or predetermined and the funds collected can be easily controlled using standard cash management techniques, and

    (b) routine voucher and invoice examination.

18. The control of the treasury accounts.

19. The administration of public trusts

APPENDIX B

The following list is of services and actions that are not considered to be inherently governmental functions. However, they may approach being in that category because of the way in which the contractor performs the contract or the manner in which the government administers contractor performance. When contracting for such services and actions, agencies should be fully aware of the terms of the contract, contractor performance, and contract administration to ensure that appropriate agency control is preserved.

This is an illustrative listing, and is not intended to promote or discourage the use of the following types of contractor services:

1. Services that involve or relate to budget preparation, including workload modeling, fact finding, efficiency studies, and should-cost analyses, etc.

2. Services that involve or relate to reorganization and planning activities.

3. Services that involve or relate to analyses, feasibility studies, and strategy options to be used by agency personnel in developing policy.

4. Services that involve or relate to the development of regulations.

5. Services that involve or relate to the evaluation of another contractor's performance.

6. Services in support of acquisition planning.

7. Contractors' providing assistance in contract management (such as where the contractor might influence official evaluations of other contractors).

8. Contractors' providing technical evaluation of contract proposals.

9. Contractors' providing assistance in the development of statements of work.

10. Contractors' providing support in preparing responses to Freedom of Information Act requests.

11. Contractors' working in any situation that permits or might permit them to gain access to confidential business information and/or any other sensitive information (other than situations covered by the Defense Industrial Security Program described in FAR 4.402(b)).

12. Contractors' providing information regarding agency policies or regulations, such as attending conferences on behalf of an agency, conducting community relations campaigns, or conducting agency training courses.

13. Contractors' participating in any situation where it might be assumed that they are agency employees or representatives.

14. Contractors' participating as technical advisors to a source selection board or participating as voting or nonvoting members of a source evaluation board.

15. Contractors' serving as arbitrators or providing alternative methods of dispute resolution.

16. Contractors' constructing buildings or structures intended to be secure from electronic eavesdropping or other penetration by foreign governments.

17. Contractors' providing inspection services.

18. Contractors' providing legal advice and interpretations of regulations and statutes to Government officials.

19. Contractors' providing special non-law enforcement, security activities that do not directly involve criminal investigations, such as prisoner detention or transport and non-military national security details.

20.

OFFICE OF MANAGEMENT AND BUDGET
Office Of Federal Procurement Policy

AGENCY:

Office of Management and Budget, Executive Office of the President, Office of Federal Procurement Policy.

ACTION:

Policy Letter on Inherently Governmental Functions.

SUMMARY:

The Office of Federal Procurement Policy (OFPP) publishes today the final version of a policy letter providing guidance to Executive Departments and agencies on (1) what functions are inherently governmental functions that must only be performed by Government officers and employees and (2) what contractible functions so closely support Government officers and employees in their performance of inherently governmental functions that the terms and performance of those contracts require closer scrutiny from Federal officials. This policy letter has been developed because executive agencies, members of Congress, the General Accounting Office, and the public have from time to time either requested guidance regarding, or inquired about, the propriety of awarding contracts for certain types of functions or administering contracts in certain ways. Previous guidance on this issue has also not been as detailed as that which we now provide..

FOR FURTHER INFORMATION CONTACT:

Richard A. Ong, Deputy Association Administrator, Office of Federal Procurement Policy, 725 17th Street, NW---Suite 9001, Washington, DC 20503 (202) 395-7209. To obtain a copy of this policy letter, please call OMB's Publication Office at (202) 395-7332.

SUPPLEMENTARY INFORMATION

Comments received. We received 34 comments in response to our proposed policy letter published in the

Federal Register on December 16, 1991 (56 Fed. Reg. 65279): eight from industry or trade groups, four from private individuals, two from employee organizations, one from a Federally funded research and development center, and 19 from Government agencies.

1. **Purpose of the policy letter**. This policy letter on inherently governmental functions is being published to provide guidance on what kinds of functions, as a matter of policy, must be performed by officials of the Executive Branch of the United States and what kinds of functions may be performed by private persons under contract with the Federal Government.

   Previous guidance on these matters that has been available to the Executive Branch has not been detailed and sometimes Federal agencies have permitted contractors to perform functions that should be performed by Government personnel. We now provide more detailed guidance.

2. **Relationships of policy letter to other OFPP publications on service contracting**. This policy letter is also one of several that the Office of Federal Procurement Policy (OFPP) has published recently that have focused on some aspect of service contracting in the Federal Government. At this time, OFPP has determined it is best to deal with individual aspects of service contracting rather than trying to publish comprehensive guidance in one document. We will consider collecting all of the guidance on service contracts in one document in the future.

   Thus, we do not cover in detail in this policy letter such matters as cost effectiveness of contracting for services, conflicts of interest of service contractors, and management of service contracts. These issues are dealt with in OMB Circular No. A-76, Performance of Commercial Activities, August 4, 1983 (under revision); OFPP Policy Letter 89-1, Conflict of interest Policies Applicable to Consultants, 54 Fed. Reg. 51,805 (December 18, 1989); OFPP Memorandum for Agency Senior Procurement Executives, Government-Wide Guidance on Contract Administration (March 15, 1991); OFPP Policy Letter 91-2, Service Contracting, 56 Fed. Reg. 15110 (April 15, 1991); proposed OFPP Policy Letter 91-___, Past Performance Information, 56 Fed. Reg. 63988 (December 6, 1991); and proposed OFPP Policy Letter 92-____, Management of Service Contracting, 56 Fed. Reg. 66091 (December 20, 1991).

3. **Relationship to OMB Circular No. A-76**. One commenter asked that we make clear our apparent intent to clarify rather than alter the guidance originally found in OMB Circular No. A-76 on inherently government functions. This is our intent. No fundamental change is intended.

   We have altered the form of the original Circular A-76 definition of an inherently governmental function in the interest of clarity. Specific examples cited in the original A-76 definition have been incorporated into Appendix A and a list of the general principles underlying the selection of the functions listed in that appendix has been added in their stead.

   The terms "function" and "activity" as used in this policy letter and Circular A-76, respectively, are interchangeable.

   The same commenter above suggested that we add a new Appendix C, containing a nonexclusive list of functions that are commercial activities hat should be contracted. We have not adopted this suggestion because the scheme proposed is the same one we have implicitly adopted. The proposed Appendix C is nothing more than the list of examples of commercial activities found as an Attachment to Circular A-76. We do not believe it is necessary to incorporate that A-76 attachment in this policy letter. The fact that we have not provided this Appendix C thus should not be construed as narrowing the scope of functions that have been contracted in the past. Nonetheless, we have added language to [[section]] 5 to clarify the relationship between Circular A-76 and this policy letter on this point.

Another commenter stated that the relationship between this policy letter and Circular A-76 is unclear. This policy letter is to be the exclusive source of guidance on what constitutes, as a matter of policy, an inherently governmental function.

4. **Libraries**. Several persons questioned the inclusion of library operations as a ministerial function that should be contracted out in subsection 7(a) of the December version of the policy letter. The fact that employees render professional services in performing a function does not mean that the function in question is necessarily inherently governmental. In fact, the Government frequently seeks out contract services precisely because of the level of sophistication required to perform a particular function. On the other hand, agencies may determine that aspects of their library operations, such as handling certain types of information in certain circumstances, involve performance of an inherently government function. Therefore, we have removed the reference to libraries.

5. **Contract audits for inspectors general**. One commenter suggested that Federal inspector general (IG) work should be done by using Government resources, with exceptions justified on a case-by-case basis, unless specific technical expertise is needed temporarily and is not available within the Government. This suggestion was not adopted because (1) Congress has specifically authorized the use of contract auditors in [[paragraph]] 6(a) (9) of the Inspector General Act codified at 5 U.S.C. App. 3, and (2) financial and compliance audit activities are not considered inherently governmental functions.

   Another commenter questioned whether subsection 12(g) of Appendix A pertaining to the determination of whether contract costs are reasonable, allocable, and allowable proscribes the use of contract audit services. It does not. The decision on what costs are reasonable, allocable, and allowable is ultimately a Government decision, but that decision may be based on recommendations made by contract auditors. Certified public accountants, for example, only render "opinions" and contracts sometimes provide that audit reports are advisory only. Moreover, the use of contract auditors has been authorized by Congress, as noted above.

6. **Agency determinations**. One commenter interpreted the policy letter as authorizing Federal managers to made a final determination on whether a function is an inherently governmental function, under this policy letter, without such determination's being subject to being overturned by the Office of Management and Budget (OMB) or being subject to a cost comparison study under Circular A-76. In general, agencies are expected to make their own determinations, subject to oversight by OMB. Language has been added to subsection 7(c) to clarify this point.

7. **Agency discretion**. One commenter questioned the need for the language in former subsection 7(e) regarding agency discretion to award nonpersonal service contracts. We agree it is unnecessary. It is already clear that awarding a contract is an agency responsibility.

8. **Incorporation in OMB Circular No. A-76, other documents**. Several commenters suggested that the policy letter be incorporated in Circular A-76, "Commercial Activities," currently being revised. We did not incorporate this suggestion because A-76 is already a lengthy document. Also, contracting for inherently government functions is indeed a consideration in contracting out, but it is not unique to the A-76 program. All Federal officials who contract for nonpersonal services must consider the problem of inherently governmental functions, and we thus believe separate guidance applicable to all such contracting, not just to nonpersonal service contracting in the A-76 context, is the better alternative. Other commenters urged that the policy letter be combined with one or more

other OFPP policy letters, such as those on conflict of interest, service contracting, and past performance and published in a form other than a policy letter. This suggestion has merit but we believe it best to try to deal with discrete portions of service contracting rather than to try to deal with all facets of a complex problem at once, as discussed in point 2, above.

9. **Agency discretion regarding resource allocation**. One commenter suggested we should address the issues of the future balance between official and contractor workforce in the performance of "basic government work," the specific expertise needed to manage the contractor workforce now or in the future, where this expertise should be located, and the way in which it can be maintained. We believe this is a matter for agencies themselves to determine, given their knowledge of their mission, their resources, the kinds of services they wish to contract, and the size of their service contracting effort. We merely highlight the problem of lack of oversight as a loss of Government control and require agencies to be aware of their existing oversight responsibilities. They are, however, to use their own discretion to figure out how to manage their contracts.

10. **Evaluation of proposals**. One commenter believes there is an apparent conflict between former subsection 14(b) in Appendix and [[section]] 8 of Appendix B. There is no conflict as new subsection 12(b) refers to participation as a voting member on source selection boards only.

11. **Appendix B controls**. The same commenter also suggested that Appendix B should contain a discussion of possible controls that the Government should employ to prevent the functions listed there from being perceived as inherently government function. We do not believe this is necessary, as any function that is in Appendix B is by definition **not** an inherently governmental function.

12. **Applicability to nonpersonal services**. Three commenters questioned why the policy letter applies only to nonpersonal service contracts. Upon consideration, we have accordingly deleted the definition of "service contract" in [[section]] 5. No useful purpose is served by defining "personal services" differently from the FAR and no harm arises from having the policy letter apply to the minimal number of true personal service contracts. Personal service contracts that are really personnel appointments are excluded from the coverage of the policy letter. Thus, FAR 37.102(b) need not be amended as a result of this policy letter.

13. **Subcontractors**. Commenters questioned whether subsection 12(d) of Appendix A should apply to subcontractors. It does not and clarifying language has been added.

14. **Supplies or services purchased by prime contractors**. Some commenters questioned the apparent effect of subsection 12 in Appendix A of preventing contractors from buying supplies and services for their own account. It is not the intent of this policy letter to prevent contractors mess halls from buying food to be prepared for military personnel. Nor does it affect what or how contractors buy to be incorporated into supplies or services to be delivered to the Government. Similarly, contractors may purchase supplies or services for the Government while acting within reasonable Government guidelines. Section 12 is only meant to address the Government's direct acquisition of supplies or services.

15. **Independent judgment**. The emphasis placed on independent judgment by this policy letter does not preclude the wholesale adoption of contractor advice, opinions, recommendations, ideas, or conclusions. They merely may not be adopted, in whole or in part, without officials' first exercising independent judgment.

16. **Duties of contracting officers**. We have added language to [[section]] 8 to spell out the analytical steps to be following by contracting officers seeking to comply with this policy letter.

17. **Risk of injury to the public**. One commenter stated that the definition of an inherently government function does not clearly address the danger to the public interest when a function is contracted out and the public is at risk if contractors, such as fire fighters or military support contractors, fail or refuse to act in time of crisis. The risk of injury to the public is an important consideration. We believe, however that [[paragraph]] 7(b)(5) appropriately identifies this point as a consideration in determining whether a function is, as a matter of policy, an inherently governmental function. The decision to include several of the functions listed in Appendix A reflects an underlying concern for this risk.

18. **Binding nature of decisions**. This same commenter noted that it is an overstatement to say that the use of discretion (referred to in what is not subsection 7(a) of the policy letter) must have the effect of committing the Government to a course of action. This is because a scientific consulting firm, for example, could submit a study that would have a tremendous impact on regulations or other agency actions but would not necessary lead to a commitment to a course of action.

    We have addressed the element of discretion in subsection 7(a) to convey the idea that the mere existence of the element of discretion is not determinative of whether, as a matter of policy, an inherently governmental function is involved. Moreover, it is useful to observe that a study hat has a tremendous impact is not per se a bad thing. A study may have that effect because of its great merit. We should be concerned, however, when a study is allowed to proceed to the point where alternative views, solutions, research, or conclusions, and so forth, cannot realistically be included or taken into account. In this case, the contractor has in effect made all important decisions. Section 7(b)(c) addresses this issue.

19. **Federally funded research and development center (FFRDCs)**. One commenter stated that while profit-making contractors can perform functions listed in Appendix B, the policy letter should cross-reference FAR 35.107 pertaining to FFRDCs and "recognize less rigorous oversight." We have not adopted this suggestion. We do not agree that FFRDCs necessarily require less oversight. FAR Part 35 and that its provisions may suffice to enable satisfactory agency oversight of FFRDCs. Whether fewer or additional control measures are necessary to ensure agency control over FFRDCs is a matter for agencies to decide in the circumstances of each case.

20. **Architect-engineer evaluation boards**. This same commenter questioned whether [[section]] 3, which states that services obtained by personnel appointments and advisory committees are not covered by this policy letter, could be construed to prohibit private individuals appointed to architect-engineer source evaluation boards in accordance with FAR 36.602 from voting. To the extent such boards are advisory committees, the policy letter is not applicable to them. If they are not, the commenter makes an excellent point. FAR 36.602-4 makes clear that the agency is to make the final selection and FAR 36.602-3(d) provides for the evaluation board to set out in its report the considerations upon which its recommendations were based. This is an acceptable mechanism and we have accordingly revised subsection 12(b) of Appendix A and [[section]] 14 of Appendix B to make clear that it is **selection** of sources that is the most sensitive function. Contractor activities that result in recommendations and that explain how those recommendations were arrived at adequately preserve agency options. A related change has been made in subsection 7(f) stating that requiring contractors to explain how they arrived at their recommendations is another available control measure.

21. Factors to consider in totality of the circumstances.

    (a) **Complexity and oversight**. One commenter questioned the inclusion of [[paragraph]] 7(d)(2) of the proposed policy letter relating to the complexity of the task to be performed. Upon consideration, we conclude that complexity is better considered in conjunction with the provision that was at 7(d)(12) relating to oversight procedures, resources, and practices. We have amended paragraph 12 accordingly and moved it, as well as the provision in former [[paragraph]] 7(d)(4) relating to the duration of the contract, to new subsection 7(e), Post-award responsibilities. This was done to remove questions relating to contract oversight from the "totality of the circumstances" test. It is important to understand that, if an agency has inadequate oversight procedures or poor oversight practices, the **underlying** function of any agency contract affected by these deficiencies is not thereby transformed into an inherently governmental function. As the totality test focuses on the nature of the function in question and as there can be a transfer of oversight responsibility even if the underlying function is contractible, the issue of de facto transfer of control should therefore be dealt with elsewhere. (Note that a transfer of contract management responsibility to the contractor is explicitly not permitted by Appendix A, subsection 12(e).)

(b) **Ultimate user of contractor work product**. Several commenters questioned the inclusion of this factor at [[paragraph]] 7(d)(3) of the proposed policy letter. We agree it should be taken out. Who will use the contractor's work product is important and this has bearing on how much management attention to give to the contract, but it doesn't say anything about the nature of the underlying function or the adequacy of agency contract administration.

(c) **Review of contractor action**. The same commenter questions the advisability of including a factor (new [[paragraph]] 7(b)(5)) that relates to the finality of any contractor's adjudication of any claim and the type of agency review of contractor adjudications. We see no problem with agencies' providing for contractor adjudication of claims so long as citizens know that they have a right of recourse to agency decisionmakers if they are dissatisfied with the decisions of the contractor. (Note, however, that certain kinds of nearings may still not be eligible of any person for a security clearance, or hearings involving actions that affect matters of personal reputation or eligibility to participate in Government programs. See Appendix A, [[section]] 14.)

Thus, we distinguish between on the one hand, holding hearings and making recommendations and, on the other, retaining the authority to issue the final adjudicatory decision. Contractors may perform the former functions so long as there is adequate oversight, agencies retain the authority to issue the final decision, and the public has a right to insist that the agency make the final decision, if it so desires. This is easier to understand if one views the contractor's action as more of a advisory action than one that binds the claimant with only limited opportunities to change the result before the agency. Note that in the absence of an appeal by a claimant, the agency need not rule on each contractor decision or ruling. It should of course, inspect or sample contractor decisions or rulings from time to time to ensure that contractors comply with agency guidelines and procedures.

(d) **Limited or extinguishing discretion**. The same commenter noted that our speaking in terms of contractor limiting or extinguishing discretion in former [[paragraph]] 7(d)(5) could mistakenly create the impression that some of the Government's authority can be exercised by a contractor. The policy letter attempts to clarify this issue at subsection 7(a).

(e) **Public perception**. Several commenters questioned the inclusion of this factor at [[paragraph]] 7(d)(11) of the proposed policy letter, believing that public perception is too ambiguous a concept. We agree. A function can probably be analyzed in the light of other factors listed without the need to resort to the concept of perceptions. Appendix A of the policy letter is itself an up-to-date listing that already takes into account the factor of public perceptions. The paragraph has been deleted.

(f) **Laws applicable to the Civil Service**. Several commenters questioned the inclusion of this factor at [[paragraph]] 7(d)(123) of the proposed policy letter. We agree and have deleted this factor. The consideration listed may be relevant to what good contract management should require by way of contract conditions, but they don't say anything about the nature of the function or the adequacy of agency contract administration practices.

(g) **Record keeping requirements**. One commenter found the meaning of paragraph 8(d)(15) of the proposed policy letter unclear. This factor was included to cover situations such as a contractor's providing a aircraft-related training. If the contractor proves to be incompetent or negligent, the fact that the contractor did maintain or was required to maintain records of who was trained permits corrective action to be taken, such as locating improperly trained students and requiring retraining. If records are not maintained, the Government cannot exercise ultimate control because it cannot correct any errors. Nonetheless, the provision appears to have only limited application and has been deleted.

22. **Collection of fees.** Two commenters questioned the provision of [[paragraph]] 20 of Appendix A of the proposed policy letter prohibiting collection of fees or other public moneys, pointing out that contractors in mess halls for military personnel currently collect charges for meals and Department of Housing and Urban Development (HUD) contractors collect fees from purchasers of HUD properties. We have modified the policy letter to enable routine collection of fees where good cash management practices and other controls are in effect, where there is little danger of miscalculating the amount of money ultimately due the Government, and where there is little difficulty in obtaining payment. For example, a contractor could have discretion to determine that a family seeking entrance to a part consists of four people rather than three, and that one of the four is a child under 12, but the contractor would not have the discretion to determine the amount of the fee to be paid by each person in a particular category. HUD contractors may also collect fees from purchasers of HUD properties in accordance with subsection 17(a) of Appendix A. We also make clear that routine voucher and invoice examination by contractors is an acceptable practice.

23. **Contract for one function or several.** One commenter questioned whether the policy letter reflects our belief that only contracts with multiple functions are susceptible to confusion with respect to inherently governmental functions. This is not our belief. The policy letter is intended to provide guidance with respect to discrete functions regardless of whether there is a mixture of several functions in a contract or there is only one function that is being contracted.

24. **Post-aware responsibilities.** Section 7(e) has been amended to make clear that agency contract oversight is to ensure contractor performance in accordance with the terms of the contract, but that oversight must not be exercised so as to create a personal service contract. Language from subsection 7(d) of the proposed policy letter has been moved to subsection 7(e), as explained in [[section]] 21, above.

25. **Drafting of Congressional testimony, responses to Congressional correspondence, and agency responses to audit reports from an Inspector General, the General Accounting Office or other Federal audit entity.** Two commenters questioned whether contractors should be able to draft Congressional testimony, subject to ultimate agency approval. Approval is a key power reserved to any official and we by no means agree that officials do or will approve contractor work in a perfunctory manner. We have nonetheless reexamined this issue and, because of the importance of Congressional testimony and correspondence of agency responses to audit reports, we are not deciding, as a matter of policy, that these documents should not be drafted by contractors. We have thus added a new subsection (c) to the body of the policy letter to this effect. We deleted the relative portions of Appendix A because we do not believe that drafting documents per se is an inherently governmental function and failing to exercise sufficient oversight with respect to drafting of such

documents does not transform the underlying function into an inherently governmental function, as noted in subsection [[paragraph]] 21(a), above. Contractor reports, conclusions, summaries, analyses, and other work products may, of course, still be quoted correspondence, and responses to audit reports, or set out in such things as attachments, appendices, or enclosures thereto.

26. **Reliance on contractor support**. One commenter called attention to our statement in [[paragraph]] 4 of the policy letter that agencies "award service contracts for various reasons, such as to acquire special skills not available in the Government or to meet the need for intermittent services." The commenter pointed out that "'support service' contractors have come to serve as the permanent workforce for many programs" seemingly implying that our statement does not take this into account. In fact, our statement is an accurate one, citing only two of the reasons why agencies award service contracts as examples. Contracting actions under Circular A-76 are also a reason why agencies award service contracts.

    Whatever the reason for using service contracts to accomplish agency missions, it is important to understand that agency use of the function must not be an inherently government function, and if it is not, the agency must be able to exercise effective oversight of any contract awarded. We make clear that management of a contract is just as important as deciding whether the contract may properly be awarded in the first place.

    Our policy letter is limited in scope and does not focus on why agencies use service contracts. Rather we are concerned that service contracts, when used, are used only when contractors may perform the functions in question and when agencies have the resources to manage the contracts. It is true that agencies have sometimes contracted functions that we have listed in the policy letter as inherently governmental functions, and it is true that they have sometimes failed to recognize that they were not exercising effective oversight over nongovernmental functions that had been contracted. Nonetheless, effective corrective action has been taken by the agencies in the past when oversight problems were identified.

    Additional problems in this area will probably arise in the future. Even the General Accounting Office recognized the difficulty in defining inherently government functions and providing guidance to agencies on the subject. **Are Service Contractors Performing Inherently Governmental Functions?**, GAO/GGD-92-11, November 1991, p. 3. We have every reason to expect, however, that because our guidance is much more detailed than anything that was available to agencies in the past there will be fewer instances of problems in this area. We thus disagree strongly with the commenter that the policy letter is a mere exhortation to better management.

27. **Other issues**. One commenter also suggested that we should address whether "contractors who perform work historically performed by civil servants should be subjected to comparable limitations on pay and rules of conduct;" measurement of the short-term and long-term costs of reliance on contractors versus officials; whether Superfund and the savings and loan bailout programs "provide models for public management of the next bailout or cleanup program;" and the "practical meaning that we will give to the concept of 'public service' as the Federal Government heads into the 21st century."

    The concept of work "historically performed" by civil servants is not useful because a function may have been performed by civil servants in the past for reasons other than the belief that the function was inherently governmental. In fact, the premise of Circular No. A-76 is that many functions historically performed by Government employees can more appropriately be performed by the private sector.

We believe that competition is the most powerful force available to keep costs down, even though there may be instances where this will not be so. In such instances, determinations shall be made in accordance with Circular No. A-76.

Measurement of the short-term and long-term costs of reliance on contractors versus officials is an aspect of cost effectiveness of service contracts and need not be dealt with here. Similarly, the efficacy of the Superfund and savings and loan programs is a matter beyond the scope of this policy letter.

So far as the practical meaning of the concept of public service is concerned, this policy letter attempts to identify those functions that, as a matter of policy, should only be performed by Government officials and those that may be performed by service contractors. If our taxonomy and analytical methods are sound, our policy letter should define what public service entails in terms of the functions that officials must perform for the foreseeable future.

28. **Acknowledgment**. Finally, we wish to acknowledge our reliance on the excellent work of the Environmental Protection Agency in our drafting of the appendices to this policy letter. Also, the comments we received were all exceptionally well thought out. We are most grateful for the time, effort and imagination that went into the preparation of those comments.

(signed)

Allan V. Burman
Administrator
Date: Sep 23 1992

PUBLIC LAW 105–270—OCT. 19, 1998

FEDERAL ACTIVITIES INVENTORY REFORM ACT OF 1998

Public Law 105–270
105th Congress

An Act

Oct. 19, 1998
[S. 314]

To provide a process for identifying the functions of the Federal Government that are not inherently governmental functions, and for other purposes.

Be it enacted by the Senate and House of Representatives of the United States of America in Congress assembled,

Federal Activities Inventory Reform Act of 1998.
31 USC 501 note.

SECTION 1. SHORT TITLE.

This Act may be cited as the "Federal Activities Inventory Reform Act of 1998".

Records.

SEC. 2. ANNUAL LISTS OF GOVERNMENT ACTIVITIES NOT INHERENTLY GOVERNMENTAL IN NATURE.

Deadline.

(a) LISTS REQUIRED.—Not later than the end of the third quarter of each fiscal year, the head of each executive agency shall submit to the Director of the Office of Management and Budget a list of activities performed by Federal Government sources for the executive agency that, in the judgment of the head of the executive agency, are not inherently governmental functions. The entry for an activity on the list shall include the following:

(1) The fiscal year for which the activity first appeared on a list prepared under this section.

(2) The number of full-time employees (or its equivalent) that are necessary for the performance of the activity by a Federal Government source.

(3) The name of a Federal Government employee responsible for the activity from whom additional information about the activity may be obtained.

(b) OMB REVIEW AND CONSULTATION.—The Director of the Office of Management and Budget shall review the executive agency's list for a fiscal year and consult with the head of the executive agency regarding the content of the final list for that fiscal year.

(c) PUBLIC AVAILABILITY OF LISTS.—

(1) PUBLICATION.—Upon the completion of the review and consultation regarding a list of an executive agency—

(A) the head of the executive agency shall promptly transmit a copy of the list to Congress and make the list available to the public; and

Federal Register, Publication.

(B) the Director of the Office of Management and Budget shall promptly publish in the Federal Register a notice that the list is available to the public.

(2) CHANGES.—If the list changes after the publication of the notice as a result of the resolution of a challenge under section 3, the head of the executive agency shall promptly—

(A) make each such change available to the public and transmit a copy of the change to Congress; and

(B) publish in the Federal Register a notice that the change is available to the public.

Federal Register, Publication.

(d) COMPETITION REQUIRED.—Within a reasonable time after the date on which a notice of the public availability of a list is published under subsection (c), the head of the executive agency concerned shall review the activities on the list. Each time that the head of the executive agency considers contracting with a private sector source for the performance of such an activity, the head of the executive agency shall use a competitive process to select the source (except as may otherwise be provided in a law other than this Act, an Executive order, regulations, or any executive branch circular setting forth requirements or guidance that is issued by competent executive authority). The Director of the Office of Management and Budget shall issue guidance for the administration of this subsection.

(e) REALISTIC AND FAIR COST COMPARISONS.—For the purpose of determining whether to contract with a source in the private sector for the performance of an executive agency activity on the list on the basis of a comparison of the costs of procuring services from such a source with the costs of performing that activity by the executive agency, the head of the executive agency shall ensure that all costs (including the costs of quality assurance, technical monitoring of the performance of such function, liability insurance, employee retirement and disability benefits, and all other overhead costs) are considered and that the costs considered are realistic and fair.

SEC. 3. CHALLENGES TO THE LIST.

(a) CHALLENGE AUTHORIZED.—An interested party may submit to an executive agency a challenge of an omission of a particular activity from, or an inclusion of a particular activity on, a list for which a notice of public availability has been published under section 2.

(b) INTERESTED PARTY DEFINED.—For the purposes of this section, the term "interested party", with respect to an activity referred to in subsection (a), means the following:

(1) A private sector source that—

(A) is an actual or prospective offeror for any contract, or other form of agreement, to perform the activity; and

(B) has a direct economic interest in performing the activity that would be adversely affected by a determination not to procure the performance of the activity from a private sector source.

(2) A representative of any business or professional association that includes within its membership private sector sources referred to in paragraph (1).

(3) An officer or employee of an organization within an executive agency that is an actual or prospective offeror to perform the activity.

(4) The head of any labor organization referred to in section 7103(a)(4) of title 5, United States Code, that includes within its membership officers or employees of an organization referred to in paragraph (3).

(c) TIME FOR SUBMISSION.—A challenge to a list shall be submitted to the executive agency concerned within 30 days after the publication of the notice of the public availability of the list under section 2.

Deadline.

(d) INITIAL DECISION.—Within 28 days after an executive agency receives a challenge, an official designated by the head of the executive agency shall—

(1) decide the challenge; and

(2) transmit to the party submitting the challenge a written notification of the decision together with a discussion of the rationale for the decision and an explanation of the party's right to appeal under subsection (e).

(e) APPEAL.—

Deadline.

(1) AUTHORIZATION OF APPEAL.—An interested party may appeal an adverse decision of the official to the head of the executive agency within 10 days after receiving a notification of the decision under subsection (d).

(2) DECISION ON APPEAL.—Within 10 days after the head of an executive agency receives an appeal of a decision under paragraph (1), the head of the executive agency shall decide the appeal and transmit to the party submitting the appeal a written notification of the decision together with a discussion of the rationale for the decision.

SEC. 4. APPLICABILITY.

(a) EXECUTIVE AGENCIES COVERED.—Except as provided in subsection (b), this Act applies to the following executive agencies:

(1) EXECUTIVE DEPARTMENT.—An executive department named in section 101 of title 5, United States Code.

(2) MILITARY DEPARTMENT.—A military department named in section 102 of title 5, United States Code.

(3) INDEPENDENT ESTABLISHMENT.—An independent establishment, as defined in section 104 of title 5, United States Code.

(b) EXCEPTIONS.—This Act does not apply to or with respect to the following:

(1) GENERAL ACCOUNTING OFFICE.—The General Accounting Office.

(2) GOVERNMENT CORPORATION.—A Government corporation or a Government controlled corporation, as those terms are defined in section 103 of title 5, United States Code.

(3) NONAPPROPRIATED FUNDS INSTRUMENTALITY.—A part of a department or agency if all of the employees of that part of the department or agency are employees referred to in section 2105(c) of title 5, United States Code.

(4) CERTAIN DEPOT-LEVEL MAINTENANCE AND REPAIR.—Depot-level maintenance and repair of the Department of Defense (as defined in section 2460 of title 10, United States Code).

SEC. 5. DEFINITIONS.

In this Act:

(1) FEDERAL GOVERNMENT SOURCE.—The term "Federal Government source", with respect to performance of an activity, means any organization within an executive agency that uses Federal Government employees to perform the activity.

(2) INHERENTLY GOVERNMENTAL FUNCTION.—

(A) DEFINITION.—The term "inherently governmental function" means a function that is so intimately related to the public interest as to require performance by Federal Government employees.

(B) FUNCTIONS INCLUDED.—The term includes activities that require either the exercise of discretion in applying Federal Government authority or the making of value judgments in making decisions for the Federal Government, including judgments relating to monetary transactions and entitlements. An inherently governmental function involves, among other things, the interpretation and execution of the laws of the United States so as—

(i) to bind the United States to take or not to take some action by contract, policy, regulation, authorization, order, or otherwise;

(ii) to determine, protect, and advance United States economic, political, territorial, property, or other interests by military or diplomatic action, civil or criminal judicial proceedings, contract management, or otherwise;

(iii) to significantly affect the life, liberty, or property of private persons;

(iv) to commission, appoint, direct, or control officers or employees of the United States; or

(v) to exert ultimate control over the acquisition, use, or disposition of the property, real or personal, tangible or intangible, of the United States, including the collection, control, or disbursement of appropriated and other Federal funds.

(C) FUNCTIONS EXCLUDED.—The term does not normally include—

(i) gathering information for or providing advice, opinions, recommendations, or ideas to Federal Government officials; or

(ii) any function that is primarily ministerial and internal in nature (such as building security, mail operations, operation of cafeterias, housekeeping, facilities operations and maintenance, warehouse operations, motor vehicle fleet management operations, or other routine electrical or mechanical services).

SEC. 6. EFFECTIVE DATE.

This Act shall take effect on October 1, 1998.

Approved October 19, 1998.

LEGISLATIVE HISTORY—S. 314:

SENATE REPORTS: No. 105–269 (Comm. on Governmental Affairs).
CONGRESSIONAL RECORD, Vol. 144 (1998):
July 30, considered and passed Senate.
Oct. 5, considered and passed House.

○

(1) Availability of purchase options.

(2) Potential for use of the equipment by other agencies after its use by the acquiring agency is ended.

(3) Trade-in or salvage value.

(4) Imputed interest.

(5) Availability of a servicing capability, especially for highly complex equipment; *e.g.*, can the equipment be serviced by the Government or other sources if it is purchased?

7.402 Acquisition methods.

(a) *Purchase method*. (1) Generally, the purchase method is appropriate if the equipment will be used beyond the point in time when cumulative leasing costs exceed the purchase costs.

(2) Agencies should not rule out the purchase method of equipment acquisition in favor of leasing merely because of the possibility that future technological advances might make the selected equipment less desirable.

(b) *Lease method.* (1) The lease method is appropriate if it is to the Government's advantage under the circumstances. The lease method may also serve as an interim measure when the circumstances—

(i) Require immediate use of equipment to meet program or system goals; but

(ii) Do not currently support acquisition by purchase.

(2) If a lease is justified, a lease with option to purchase is preferable.

(3) Generally, a long term lease should be avoided, but may be appropriate if an option to purchase or other favorable terms are included.

(4) If a lease with option to purchase is used, the contract shall state the purchase price or provide a formula which shows how the purchase price will be established at the time of purchase.

7.403 General Services Administration assistance.

(a) When requested by an agency, the General Services Administration (GSA) will assist in lease or purchase decisions by providing information such as—

(1) Pending price adjustments to Federal Supply Schedule contracts;

(2) Recent or imminent technological developments;

(3) New techniques; and

(4) Industry or market trends.

(b) Agencies may request information from the following GSA offices:

(1) Center for Strategic IT Analysis (MKS), Washington, DC 20405, for information on acquisition of information technology.

(2) Federal Supply Service, Office of Acquisition (FC), Washington, DC 20406, for information on other types of equipment.

7.404 Contract clause.

The contracting officer shall insert a clause substantially the same as the clause in 52.207-5, Option to Purchase Equipment, in solicitations and contracts involving a lease with option to purchase.

Subpart 7.5—Inherently Governmental Functions

7.500 Scope of subpart.

The purpose of this subpart is to prescribe policies and procedures to ensure that inherently governmental functions are not performed by contractors. It implements the policies of Office of Federal Procurement Policy (OFPP) Policy Letter 92-1, Inherently Governmental Functions.

7.501 Definition.

"Inherently governmental function" means, as a matter of policy, a function that is so intimately related to the public interest as to mandate performance by Government employees. This definition is a policy determination, not a legal determination. An inherently governmental function includes activities that require either the exercise of discretion in applying Government authority, or the making of value judgments in making decisions for the Government. Governmental functions normally fall into two categories: the act of governing, *i.e.*, the discretionary exercise of Government authority, and monetary transactions and entitlements.

(a) An inherently governmental function involves, among other things, the interpretation and execution of the laws of the United States so as to—

(1) Bind the United States to take or not to take some action by contract, policy, regulation, authorization, order, or otherwise;

(2) Determine, protect, and advance United States economic, political, territorial, property, or other interests by military or diplomatic action, civil or criminal judicial proceedings, contract management, or otherwise;

(3) Significantly affect the life, liberty, or property of private persons;

(4) Commission, appoint, direct, or control officers or employees of the United States; or

(5) Exert ultimate control over the acquisition, use, or disposition of the property, real or personal, tangible or intangible, of the United States, including the collection, control, or disbursement of Federal funds.

(b) Inherently governmental functions do not normally include gathering information for or providing advice, opinions, recommendations, or ideas to Government officials. They also do not include functions that are primarily ministerial and internal in nature, such as building security, mail operations, operation of cafeterias, housekeeping, facilities operations and maintenance, warehouse operations, motor vehicle fleet management operations, or other routine electrical or mechanical services. The list of commercial activities included in the attachment to Office of Management and Budget (OMB) Circular No. A-76 is an authoritative, nonexclusive list of functions which are not inherently governmental functions.

7.502 Applicability.

The requirements of this subpart apply to all contracts for services. This subpart does not apply to services obtained through either personnel appointments, advisory committees, or personal services contracts issued under statutory authority.

7.503 Policy.

(a) Contracts shall not be used for the performance of inherently governmental functions.

(b) Agency decisions which determine whether a function is or is not an inherently governmental function may be reviewed and modified by appropriate Office of Management and Budget officials.

(c) The following is a list of examples of functions considered to be inherently governmental functions or which shall be treated as such. This list is not all inclusive:

(1) The direct conduct of criminal investigations.

(2) The control of prosecutions and performance of adjudicatory functions other than those relating to arbitration or other methods of alternative dispute resolution.

(3) The command of military forces, especially the leadership of military personnel who are members of the combat, combat support, or combat service support role.

(4) The conduct of foreign relations and the determination of foreign policy.

(5) The determination of agency policy, such as determining the content and application of regulations, among other things.

(6) The determination of Federal program priorities for budget requests.

(7) The direction and control of Federal employees.

(8) The direction and control of intelligence and counter-intelligence operations.

(9) The selection or non-selection of individuals for Federal Government employment, including the interviewing of individuals for employment.

(10) The approval of position descriptions and performance standards for Federal employees.

(11) The determination of what Government property is to be disposed of and on what terms (although an agency may give contractors authority to dispose of property at prices within specified ranges and subject to other reasonable conditions deemed appropriate by the agency).

(12) In Federal procurement activities with respect to prime contracts—

(i) Determining what supplies or services are to be acquired by the Government (although an agency may give contractors authority to acquire supplies at prices within specified ranges and subject to other reasonable conditions deemed appropriate by the agency);

(ii) Participating as a voting member on any source selection boards;

(iii) Approving any contractual documents, to include documents defining requirements, incentive plans, and evaluation criteria;

(iv) Awarding contracts;

(v) Administering contracts (including ordering changes in contract performance or contract quantities, taking action based on evaluations of contractor performance, and accepting or rejecting contractor products or services);

(vi) Terminating contracts;

(vii) Determining whether contract costs are reasonable, allocable, and allowable; and

(viii) Participating as a voting member on performance evaluation boards.

(13) The approval of agency responses to Freedom of Information Act requests (other than routine responses that, because of statute, regulation, or agency policy, do not require the exercise of judgment in determining whether documents are to be released or withheld), and the approval of agency responses to the administrative appeals of denials of Freedom of Information Act requests.

(14) The conduct of administrative hearings to determine the eligibility of any person for a security clearance, or involving actions that affect matters of personal reputation or eligibility to participate in Government programs.

(15) The approval of Federal licensing actions and inspections.

(16) The determination of budget policy, guidance, and strategy.

(17) The collection, control, and disbursement of fees, royalties, duties, fines, taxes, and other public funds, unless authorized by statute, such as 31 U.S.C. 952 (relating to private collection contractors) and 31 U.S.C. 3718 (relating to private attorney collection services), but not including—

(i) Collection of fees, fines, penalties, costs, or other charges from visitors to or patrons of mess halls, post or base exchange concessions, national parks, and similar entities or activities, or from other persons, where the amount to be collected is easily calculated or predetermined and the funds collected can be easily controlled using standard case management techniques; and

(ii) Routine voucher and invoice examination.

(18) The control of the treasury accounts.

(19) The administration of public trusts.

(20) The drafting of Congressional testimony, responses to Congressional correspondence, or agency responses to audit reports from the Inspector General, the General Accounting Office, or other Federal audit entity.

(d) The following is a list of examples of functions generally not considered to be inherently governmental functions. However, certain services and actions that are not considered to be inherently governmental functions may approach being in that category because of the nature of the function, the manner in which the contractor performs the contract, or the manner in which the Government administers contractor performance. This list is not all inclusive:

(1) Services that involve or relate to budget preparation, including workload modeling, fact finding, efficiency studies, and should-cost analyses, etc.

(2) Services that involve or relate to reorganization and planning activities.

(3) Services that involve or relate to analyses, feasibility studies, and strategy options to be used by agency personnel in developing policy.

(4) Services that involve or relate to the development of regulations.

(5) Services that involve or relate to the evaluation of another contractor's performance.

(6) Services in support of acquisition planning.

(7) Contractors providing assistance in contract management (such as where the contractor might influence official evaluations of other contractors).

(8) Contractors providing technical evaluation of contract proposals.

(9) Contractors providing assistance in the development of statements of work.

(10) Contractors providing support in preparing responses to Freedom of Information Act requests.

(11) Contractors working in any situation that permits or might permit them to gain access to confidential business information and/or any other sensitive information (other than situations covered by the National Industrial Security Program described in 4.402(b)).

(12) Contractors providing information regarding agency policies or regulations, such as attending conferences on behalf of an agency, conducting community relations campaigns, or conducting agency training courses.

(13) Contractors participating in any situation where it might be assumed that they are agency employees or representatives.

(14) Contractors participating as technical advisors to a source selection board or participating as voting or nonvoting members of a source evaluation board.

(15) Contractors serving as arbitrators or providing alternative methods of dispute resolution.

(16) Contractors constructing buildings or structures intended to be secure from electronic eavesdropping or other penetration by foreign governments.

(17) Contractors providing inspection services.

(18) Contractors providing legal advice and interpretations of regulations and statutes to Government officials.

(19) Contractors providing special non-law enforcement, security activities that do not directly involve criminal investigations, such as prisoner detention or transport and non-military national security details.

(e) Agency implementation shall include procedures requiring the agency head or designated requirements official to provide the contracting officer, concurrent with transmittal of the statement of work (or any modification thereof), a written determination that none of the functions to be performed are inherently governmental. This assessment should place emphasis on the degree to which conditions and facts restrict the discretionary authority, decision-making responsibility, or accountability of Government officials using contractor services or work products. Disagreements regarding the determination will be resolved in accordance with agency procedures before issuance of a solicitation.

* * * * * *

Appendix D

Questionnaire

APPENDIX D

Questionnaire

NATIONAL RESEARCH COUNCIL
COMMISSION ON ENGINEERING AND TECHNICAL SYSTEMS
BOARD ON INFRASTRUCTURE AND THE CONSTRUCTED ENVIRONMENT

Mailing Address:
2101 Constitution Avenue
Washington, DC 20418
FAX: (202)334-3370

Office Location:
Harris Building Room 274
2001 Wisconsin Avenue, N.W.
(202)334-3374

July 30, 1997

name
title
address
city, state zip
phone

name

For many years, Federal agencies have outsourced building design and construction activities while retaining management responsibilities for these activities in-house. With the downsizing of agencies and in response to legislative initiatives, more agencies are moving towards the outsourcing of **the management of** their planning, design and construction related services. At the request of the Federal Facilities Council, the National Research Council has established a committee to develop a decision model for federal agencies to use in determining whether and under what circumstances to outsource the management of planning, design, and construction related services. The study committee is chaired by Mr. Henry Michel, Chairman Emeritus of Parsons, Brinckerhoff Inc., and includes other distinguished experts (Enclosure 1). This study will also: 1) assess recent federal experience with outsourcing of the management of planning, design, and construction related services; 2) develop a technical framework and methodology for successfully implementing such an outsourcing program; 3) identify measures to determine performance outcomes; and, 4) identify the organizational core competencies needed to effectively oversee the management of outsourcing contracts and ensure protection of the federal interest.

To better understand federal agencies' expectations for and experience with the outsourcing of the management of planning, design and construction related services, the Committee is asking that sponsoring agencies of the Federal Facilities Council and other selected agencies fill out the attached questionnaire (Enclosure 2).

It would be greatly appreciated if you or an appropriate designee could return the questionnaire to me on or before **Friday, September 5, 1997**. Completed questionnaires may be faxed to (202) 334-3370, or mailed to the National Research Council, 2101 Constitution Avenue NW, HA-274,

Washington, DC 20418. If you would like to receive a copy of the questionnaire by electronic mail please send a note to jwalewsk@nas.edu. Please call (202)334-3384 if you have any questions.

Cordially,

John Walewski
Board on Infrastructure and the
Constructed Environment

BOARD ON INFRASTRUCTURE AND THE CONSTRUCTED ENVIRONMENT
Committee on the Outsourcing of Planning, Design, and Construction Related Services for Federal Facilities
Questionnaire

For many years, Federal agencies have outsourced building design and construction activities while retaining management responsibilities for these activities in-house. With the downsizing of agencies and in response to legislative initiatives, more agencies are moving towards the outsourcing of **the management of** their planning, design and construction related services. In response to a request by the Federal Facilities Council, the Board on Infrastructure and the Constructed Environment of the National Research Council has established a committee to develop a decision model for federal agencies to use in determining whether and under what circumstances to outsource the management of planning, design, and construction related services. The study will also: 1) assess recent federal experience with outsourcing of the management of planning, design, and construction related services; 2) develop a technical framework and methodology for successfully implementing such an outsourcing program; 3) identify measures to determine performance outcomes; and 4) identify the organizational core competencies needed to provide effective oversight and protect the federal interest.

The attached questionnaire is designed to gather background information for use by the Committee in formulating its recommendations. Questions #1-18 are general and relate to the management of planning, design, and construction related services within your organization and expectations related to the outsourcing of the management of such services. To the extent possible, all agencies should respond. Questions #19-28 should only be answered by agencies with actual experience in the **outsourcing of the management** of planning, design, and construction-related services. Comments, additional information and detailed clarifications are welcome.

Please return the questionnaire to Mr. John Walewski, Program Officer, Board on Infrastructure and the Constructed Environment, on or before **Friday, September 5, 1997**. The questionnaire may be faxed (202) 334-3370 or mailed to: 2101 Constitution Avenue, NW, HA-274, Washington, DC 20418. If you would like to receive an electronic version of the questionnaire, please send a note to jwalewsk@nas.edu. If you have any questions concerning the questionnaire or the current study, please call (202) 334-3384.

Name of organization ______________________________

____

Point of contact ______________________________

Mailing address ______________________________

____

Telephone number ____________________ Facsimile number ______________

E-mail address ____________________

1. Name of the office within your organization that is responsible for the management of planning, design and construction related services (example: Office of Planning and Construction Services). *note: if more than one office, please provide names for each.*

____
____
____
____

2. What was your agency's budget for planning, design, and construction related services in Fiscal Year 1996 (in millions of dollars?)

____
____
____

3. Does your agency have a generic program/project flow diagram or list of activities it uses for acquiring new facilities, including the identification and justification of funding? ________ If yes, could you attach a copy, including any explanatory information?

____
____
____

4. How is the process of program/project implementation in your agency currently managed? Please be as specific as possible regarding staff responsibilities, project monitoring, etc.

____
____
____
____
____
____
____
____
____
____
____

5. Where in this implementation process does, or could your agency's responsibility for the outsourcing of the management of planning, design, and construction lie?

____
____
____
____
____
____
____
____

6. Office of Federal Procurement Policy Letter 92-1 outlines the inherent functions and responsibilities of government. Has your agency developed additional guidance regarding the inherent functions of government? __________ If yes, please explain briefly and provide a copy, if possible.

7. Does your organization have a written policy or guidelines related to the outsourcing of planning, design or construction services and/or the outsourcing of the management of such services? _________ If yes, please explain briefly and provide a copy if possible.

8. Has your organization ever outsourced for planning, design or construction related services while retaining management responsibilities in-house? __________ If yes, for how many years? __________

9. Was your agency's work outsourced to other federal agencies or to the private sector?

10. How did your internal "end-users" or customers interface with your outsourcing contractors?

11. Under what conditions and circumstances do you feel outsourcing of the management of planning, design and construction may be appropriate for your agency?

12. What do you see as the potential barriers to successfully implementing a program to outsource the management of planning, design and construction related services? Please check all that apply and explain below.

- ☐ Federal Acquisition Regulations?
- ☐ Internal Agency/Enterprise Skills?
- ☐ Cultural/Institutional Constraints?
- ☐ Internal Systems and Controls?
- ☐ Other.

13. What do you see as your agency's potential risk in outsourcing the management of planning, design, and construction-related activities?

14. What do you see as the potential for "added value" for your agency in outsourcing the management of planning, design, and construction related services?

15. What do you see as the potential impacts of the outsourcing of the management of planning, design and construction related services on the quality of services received?

16. What performance measures do you feel would be important to look at in order to fully evaluate the results of a program to outsource the management of planning, design and construction services?

____
____
____
____

17. What would you identify as key elements for successfully implementing a program to outsource the management of planning, design, and construction-related services?

____
____
____
____
____
____
____
____
____

18. Has your organization ever outsourced for the **management** of planning, design or construction related services? _________ If no, stop here. If yes, please answer all of the following questions.

19. Please identify the reasons your agency determined it would outsource the management of planning, design, and construction related services. Check all that apply, asterisk the **most** important reason.

   - ❑ A means of reducing costs
   - ❑ Savings on project delivery time
   - ❑ A means of improving quality of product
   - ❑ Lack of in-house expertise
   - ❑ Staff shortage
   - ❑ Deliberate downsizing decision
   - ❑ Other, please explain. ____

   ____
   ____

20. For how many years has your agency been outsourcing of the management of planning, design, and construction-related services been used? __________ For approximately how many projects? ____

____
____

21. In general, for what type and value of planning, design, and construction activities was the management outsourced?

22. How were the various responsibilities related to the management of planning design, and construction related services defined for the contractor?

23. What was the process for validating and controlling any project changes and cost growth?

24. Was accountability to each of the parties detailed and clearly understood? ___________
What means were enacted to assure proper management control?

25. Were performance standards established or other methods to measure achievements initiated? ___________ If so, what were they?

26. What were the key outcomes (program/projects) of outsourcing the managment of planning, design, and construction related services?

____
____
____
____
____
____
____
____
____

27. Have any studies been made to compare the relative cost of outsourcing the management of planning, design, or construction related services versus performing it in-house? ________ Have any of these studies been validated and by whom?

____
____
____
____
____
____
____

28. Has your organization compiled any before (i.e. A-76 study) or after-action analyses which outline the decision methodology, the processes employed, the results obtained and the benefits and shortcomings exposed with outsourcing the management of planning, design, and construction related services? ______ If yes, could you provide a copy of the report(s)?

____
____
____
____

Please attach documents or include any comments which you believe would aide the committee in its effort.

Thank you for your time and assistance.

Bibliography

Air Force Logistics Management Agency. 1996. Outsourcing Guide for Contracting. Project Number LC9608100. Maxwell Air Force Base, Alabama: Air Force Logistics Management Agency.

American Society of Civil Engineers. 1990. Quality in the Constructed Project: A Guide for Owners, Designers and Constructors. Manuals and Reports on Engineering Practice No. 73. New York: American Society of Civil Engineers.

1996. Government Relations: Educational and Training Needs of Government Engineers. ASCE Policy Statement 394. New York: American Society of Civil Engineers.

1996. Government Relations: Recruitment and Retention of Qualified Engineers for Government Service. ASCE Policy Statement. New York: American Society of Civil Engineers.

Arizona Department of Transportation. 1995. Policies and Procedures: ADOT Competitive Government Services Policy. MGT-2.01. Phoenix: Arizona Department of Transportation.

Bachner, J.P. 1996. Purchasing the Services of Engineers, Architects, and Environmental Professionals. Silver Spring, Md.: Quixote Press.

Boroush, M. 1998. Understanding Risk Analysis: A Short Guide for Health, Safety, and Environmental Policy Making. Washington, D.C.: American Chemical Society.

Boston, J. 1995. Inherently Governmental Functions and the Limits to Contracting Out. Pp. 78–111 in The State Under Contract, J. Boston ed. Wellington, New Zealand: Bridget Williams Books Ltd.

Business Roundtable. 1997. The Business Stake in Effective Project Systems. Washington, D.C.: The Business Roundtable, Construction Cost Effectiveness Task Force.

Camm, F. 1996. Expanding Private Production of Defense Services. Prepared for the Commission on Roles and Missions of the Armed Forces, National Defense Research Institute. Washington, D.C.: RAND.

Center for Construction Industry Studies. 1999. Owner/Contractor Organizational Changes. Phase II Report. Austin, Texas: Center for Construction Industry Studies, University of Texas Press.

Construction Industry Institute. 1986. Evaluation of Design Effectiveness. Report RS8-1. Austin, Texas: The University of Texas Press.

Construction Industry Institute. 1996. Owner/ Contractor Work Structure: A Preview. Owner/ Contractor Work Structure Research Team, The Construction Industry Institute. Austin, Texas: The Construction Industry Institute.

Construction Management Association of America. 1993. An Introduction to Professional Construction Management. McLean, Va.: Construction Management Association of America, Inc.
Deloitte & Touche LLP. 1997. Impact of Reengineering on Corporate Real Estate. A Deloitte & Touche/NACORE Benchmarking Study. Chicago, Ill.: Deloitte & Touche.
Department of the Army. 1996. Organization and Functions, U.S. Army Corps of Engineers, Division and District Offices. ER 10-1-2. Washington, D.C.: U.S. Army Corps of Engineers.
Department of the Army. 1998. Competitive Sourcing Strategic Plan for the U.S. Army. Washington, D.C.: Department of the Army: Office of the Assistant Chief of Staff for Installation Management.
Dilger, R.J., R.R. Moffett, and L. Struyk. 1997. Privatization of municipal services in America's largest cities. Public Administration Review 57(1): 21–26.
Executive Office of the President. 1983. Performance of Commercial Activities Circular No. A-76, Office of Management and Budget, Executive Office of the President. Washington, D.C.: Office of Management and Budget.
Federal Construction Council. 1987. Quality Control on Federal Construction Projects. Technical Report No. 84. Consulting Committee on Contract Management. Washington, D.C.: National Academy Press.
Federal Facilities Council. 1998. Government/Industry Forum on Capital Facilities and Core Competencies. Technical Report #136. Washington, D.C.: National Academy Press.
Federal Facilities Council. 2000. Adding Value to the Facility Acquisition Process: Best Practices for Reviewing Facility Designs. Washington, D.C.: National Academy Press.
Gallon, M.R., H.M. Stillman, and D. Coates. 1998. Putting Core Competency Thinking into Practice. Available on line at: *http://www.iriinc.org/hot1.htm*
Gateway Group. 1998. Performance Improvement through Quality Management. Available on line at: *http://www.pw.com/us/gsa-qm.htm*
General Accounting Office (GAO). 1992. Government Contractors: Are Service Contractors Performing Inherently Governmental Functions? Report to the Chairman, Federal Service, Post Office and Civil Service Subcommittee, Committee on Governmental Affairs, U.S. Senate. Washington, D.C.: Government Printing Office.
GAO. 1994. Budget Issues: Budget Scorekeeping for Acquisition of Federal Buildings. T-AIMD-94-189. Washington, D.C.: Government Printing Office.
GAO. 1994. Contract Pricing: DOD Management of Contractors with High-Risk Cost Estimating Systems. NSIAD-94-153. Washington, D.C.: Government Printing Office.
GAO. 1994. Contractor Overhead Costs: Money Saving Reviews Are Not Being Done as Directed. Report to the Secretary of Defense. NSIAD-94-205. Washington, D.C.: Government Printing Office.
GAO. 1994. Defense Budget: Capital Asset Projects Undergo Significant Change between Approval and Execution. Report to Congressional Requesters. NSIAD-95-20. Washington, D.C.: Government Printing Office.
GAO. 1994. Defense Management Initiatives: Limited Progress in Implementing Management Initiatives. T-AIMD-94-105. Washington, D.C.: Government Printing Office.
GAO. 1994. Government Contractors: Contracting Out Implications of Streamlining Agency Operations. T-GGD-95-4. Washington, D.C.: Government Printing Office.
GAO. 1994. Government Contractors: Measuring Costs of Service Contractors versus Federal Employees. Report to the Chairman, Subcommittee on Federal Services, Post Office and Civil Service, Committee on Governmental Affairs, U.S. Senate. GGD-94-95. Washington, D.C.: Government Printing Office.
GAO. 1994. Improving Government: Actions Needed to Sustain and Enhance Management Reforms. T-OCG-94-1. Washington, D.C.: Government Printing Office.

GAO. 1994. NASA Procurement: Contract and Management Improvements at the Jet Propulsion Laboratory. Report to Congressional Requesters. NSIAD-95-40. Washington, D.C.: Government Printing Office.

GAO. 1994. Overhead Costs: Unallowable and Questionable Costs Charged by Government Contractors.T-NSIAD-94-132. Washington, D.C.: Government Printing Office.

GAO. 1994. Public-Private Mix: Extent of Contracting Out for Real Property Management Services in GSA. Briefing Report to the Ranking Minority Member, Subcommittee on Investigations and Oversight, Committee on Public Works and Transportation, House of Representatives. GGD-94-126BR. Washington, D.C.: Government Printing Office.

GAO. 1995. Government Contractors: An Overview of the Federal Contracting-Out Program. T-GGD-95-131. Washington, D.C.: Government Printing Office.

GAO. 1995. Government Reform: Using Reengineering and Technology to Improve Government Performance. T-OCG-95-2. Washington, D.C.: Government Printing Office.

GAO. 1995. Government Reorganization: Issues and Principles. T-GGD/AIMD-95-166. Washington, D.C.: Government Printing Office.

GAO. 1995. Managing for Results: Critical Actions for Measuring Performance. T-GGD/AIMD-95-187. Washington, D.C.: Government Printing Office.

GAO. 1995. Public-Private Mix: Effectiveness and Performance of GSA's In-House and Contracted Services. National Performance Review. GGD-95-204. Washington, D.C.: Government Printing Office.

GAO. 1996. Acquisition Reform: Efforts to Reduce the Cost to Manage and Oversee DOD Contracts. Report to Congressional Committees. NSIAD-96-106. Washington, D.C.: Government Printing Office.

GAO. 1996. Airport Privatization: Issues Related to the Sale or Lease of U.S. Commercial Airports. Report to the Subcommittee on Aviation, Committee on Transportation and Infrastructure, House of Representatives. RCED-97-3. Washington, D.C.: Government Printing Office.

GAO. 1996. DOD Training: Opportunities Exist to Reduce the Training Infrastructure. Report to Congressional Committees. NSIAD-96-93. Washington, D.C.: Government Printing Office.

GAO. 1996. Energy Department Trade Missions: Authority, Results, and Management Issues. T-NSIAD-96-151. Washington, D.C.: Government Printing Office.

GAO. 1996. Federal Contracting: Comments on S. 1724, the Freedom from Government Competition Act. Testimony before the Committee on Governmental Affairs, U.S. Senate. T-GGD-96-169. Washington, D.C.: Government Printing Office.

GAO. 1997. Base Operations: Challenges Confronting DOD as It Renews Emphasis on Outsourcing. Report to the Chairman, Subcommittee on Military Readiness, Committee on National Security, House of Representatives. NSIAD-97-86. Washington, D.C.: Government Printing Office.

GAO. 1997. Defense Acquisition Organizations: Linking Workforce Reductions with Better Program Outcomes. Testimony before the Subcommittees on Military Procurement and Military Readiness, Committee on National Security, House of Representatives. T-NSIAD-97-140. Washington, D.C.: Government Printing Office.

GAO. 1997. Defense Outsourcing: Challenges Facing DOD as It Attempts to Save Billions in Infrastructure Costs. Testimony Before the Subcommittee on Readiness, Committee on National Security, House of Representatives. T-NSIAD-97-110. Washington, D.C.: Government Printing Office.

GAO. 1997. Department of Energy: Clearer Missions and Better Management Are Needed at the National Laboratories. Testimony before the Subcommittee on Oversight and Investigations, Committee on Commerce, House of Representatives. T-RCED-98-25. Washington, D.C.: Government Printing Office.

GAO. 1997. Department of Energy Contract Management. High-Risk Series. HR-97-13. Washington, D.C.: Government Printing Office.

GAO. 1997. Department of Energy: Improving Management of Major System Acquisitions. Testimony before the Subcommittee on Energy and Environment, Committee on Science, House of Representatives. T-RCED-97-92. Washington, D.C.: Government Printing Office.

GAO. 1997. Federal Management and Workforce Issue Area Plan, Fiscal Year 1997. IAP-96-15. Washington, D.C.: Government Printing Office.

GAO. 1997. Federal Management and Workforce Issues Area Plan, Fiscal Years 1997-98. IAP-97-5. Washington, D.C.: Government Printing Office.

GAO. 1997. Outsourcing DOD Logistics: Savings Achievable But Defense Science Board's Projections Are Overstated. Report to Congressional Requesters. NSIAD-98-48. Washington, D.C.: Government Printing Office.

GAO. 1997. Privatization and Competition: Comments on H.R. 716, the Freedom from Government Competition Act. Testimony before the Committee on Government Management, Information and Technology, Committee on Government Reform and Oversight, House of Representatives. T-GGD-97-185. Washington, D.C.: Government Printing Office.

GAO. 1997. Privatization and Competition: Comments on S. 314, the Freedom from Government Competition Act. Testimony before the Subcommittee on Oversight of Government Management, Restructuring, and the District of Columbia, Committee on Government Affairs, United States Senate. T-GGD-97-134. Washington, D.C.: Government Printing Office.

GAO. 1997. Privatization: Lessons Learned by State and Local Governments. Report to the Chairman, House Republican Task Force on Privatization. GGD-97-48. Washington, D.C.: Government Printing Office.

GAO. 1997. Space Station: Deteriorating Cost and Schedule Performance under the Prime Contract. Testimony before the Subcommittee on Science, Technology, and Space, Committee on Commerce, Science, and Transportation, U.S. Senate. T-NSIAD-97-262. Washington, D.C.: Government Printing Office.

GAO. 1997. Terms Related to Privatization Activities and Processes: Glossary. GGD-97-121. Washington, D.C.: Government Printing Office.

GAO. 1998. Base Operations: DOD's Use of Single Contracts for Multiple Support Services. Report to Congressional Committees. NSIAD-98-82. Washington, D.C.: Government Printing Office.

GAO. 1998. Defense Outsourcing: Better Data Needed to Support Overhead Rates for A-76 Studies. Report to the Honorable Henry Bonilla, House of Representatives. NSIAD-98-62. Washington, D.C.: Government Printing Office.

GAO. 1998. Defense Outsourcing: Impact on Navy Sea-Shore Rotations. Report to Congressional Requesters. NSIAD-98-107. Washington, D.C.: Government Printing Office.

GAO. 1998. Environmental Protection: EPA's and States' Efforts to Focus State Enforcement Programs on Results. Testimony before the Subcommittee on Oversight and Investigations, Committee on Commerce, House of Representatives. T-RCED-98-233. Washington, D.C.: Government Printing Office.

GAO. 1998. Executive Guide: Leading Practices in Capital Decision-Making. AIMD-99-32. Washington, D.C.: Government Printing Office.

GAO. 1998. Federal Management Issues. Report to Congressional Requesters. OCG-98-1R. Washington, D.C.: Government Printing Office.

GAO. 1998. OMB Circular A-76: Oversight and Implementation Issues. Testimony before the Subcommittee on Oversight of Government Management, Restructuring and the District of Columbia, Committee on Government Affairs, United States Senate. T-GGD-98-146. Washington, D.C.: Government Printing Office.

GAO. 1999. Agency Performance Plans: Examples of Practices That Can Improve Usefulness to Decisionmakers. Report to the Chairman, Committee on Governmental Affairs, U.S. Senate. GGD/AIMD-99-69. Washington, D.C.: Government Printing Office.

GAO. 1999. Coast Guard: Key Budget Issues for Fiscal Years 1999 and 2000. Testimony before the Subcommittee on Coast Guard and Maritime Transportation, Committee on Transportation and Infrastructure, House of Representatives. T-RCED-99-83. Washington, D.C.: Government Printing Office.

GAO. 1999. Coast Guard: Strategies for Procuring New Ships, Aircraft, and Other Assets. Testimony before the Subcommittee on Transportation and Related Agencies, Committee on Appropriations, House of Representatives. T-RCED-99-116. Washington, D.C.: Government Printing Office.

GAO. 1999. Contract Management: DOD Pricing of Commercial Items Needs Continued Emphasis. Report to Congressional Requesters, NSIAD-99-90. Washington, D.C.: Government Printing Office.

GAO. 1999. Defense Acquisition: Best Commercial Practices Can Improve Program Outcomes. Testimony before the Subcommittee on Readiness and Management Support, Committee on Armed Services, U.S. Senate. T-NSIAD-99-116. Washington, D.C.: Government Printing Office.

GAO. 1999. Defense Infrastructure: Improved Performance Measures Would Enhance Defense Reform Initiative. Report to the Chairman, Committee on Armed Services, House of Representatives. NSIAD-99-169. Washington, D.C.: Government Printing Office.

GAO. 1999. Department of Energy: Challenges Exist in Managing the Spallation Neutron Source Project. Testimony before the Subcommittee on Energy and Environment, Committee on Science, House of Representatives. T-RCED-99-103. Washington, D.C.: Government Printing Office.

GAO. 1999. Department of Energy: Need to Address Longstanding Management Weaknesses. Testimony before the Subcommittee on Energy and Environment, Committee on Science and the Subcommittee on Energy and Power, Committee on Commerce, House of Representatives. T-RCED-99-255. Washington, D.C.: Government Printing Office.

GAO. 1999. DOD Competitive Sourcing: Lessons Learned System Could Enhance A-76 Study Process. Report to Congressional Committees. NSIAD-99-152. Washington, D.C.: Government Printing Office.

GAO. 1999. DOD Competitive Sourcing: Questions about Goals, Pace, and Risks of Key Reform Initiative. Report to the Chairman, Subcommittee on Military Readiness, Committee on Armed Services, House of Representatives. NSIAD-99-46. Washington, D.C.: Government Printing Office.

GAO. 1999. DOD Competitive Sourcing: Results of Recent Competitions. Report to the Chairman, Subcommittee on Readiness and Management Support, Committee on Armed Services, U.S. Senate. NSIAD-99-44. Washington, D.C.: Government Printing Office.

GAO. 1999. Federal Management: Challenges Facing the Department of Transportation. Testimony before the Subcommittee on Transportation, Committee on Appropriations, U.S. Senate. T-RCED/AIMD-99-94. Washington, D.C.: Government Printing Office.

GAO. 1999. Federal Workforce: Payroll and Human Capital Changes during Downsizing. Report to Congressional Requesters. GGD-99-57. Washington, D.C.: Government Printing Office.

GAO. 1999. Financial Management: Better Controls Essential to Improve the Reliability of DOD's Depot Inventory Records. Report to Agency Officials. AIMD-99-132. Washington, D.C.: Government Printing Office.

GAO. 1999. Force Structure: A-76 Not Applicable to Air Force 38th Engineering Installation Wing Plan. Report to Congressional Requesters. NSIAD-99-73. Washington, D.C.: Government Printing Office.

GAO. 1999. Future Years Defense Program: How Savings from Reform Initiatives Affect DOD's 1999–2003 Program. Report to the Chairman and Ranking Minority Member, Committee on Armed Services, House of Representatives. NSIAD-99-69. Washington, D.C.: Government Printing Office.

GAO. 1999. Government Management: Addressing High Risks and Improving Performance and Accountability. Testimony before the Committee on Government Reform, House of Representatives. T-OCG-99-23. Washington, D.C.: Government Printing Office.

GAO. 1999. Human Capital: A Self-Assessment Checklist for Agency Leaders. GGD-99-179. Washington, D.C.: Government Printing Office.

GAO. 1999. Major Management Challenges and Program Risks: A Governmentwide Perspective. OCG-99-1. Washington, D.C.: Government Printing Office.

GAO. 1999. Major Management Challenges and Program Risks: Department of Defense. OCG-99-4. Washington, D.C.: Government Printing Office.

GAO. 1999. Major Management Challenges and Program Risks: Department of the Interior. OCG-99-9. Washington, D.C.: Government Printing Office.

GAO. 1999. Major Management Challenges and Program Risks: National Aeronautics and Space Administration. OCG-99-18. Washington, D.C.: Government Printing Office.

GAO. 1999. National Laboratories: DOE Needs to Assess the Impact of Using Performance-Based Contracts. Report to the Committee on Science, House of Representatives. RCED-99-141. Washington, D.C.: Government Printing Office.

GAO. 1999. Public-Private Partnerships: Terms Related to Building and Facility Partnerships: Glossary. GGD-99-71. Washington, D.C.: Government Printing Office

General Services Administration. 1998. Design and Construction Delivery Process. Available on line at: *http://www.gsa.gov/pbs/pc/ds_files/delivery.htm#overview.*

General Services Administration. 1998. Outsourcing Information Technology. White Paper. IT Management Practices Division, Office of Information Technology, Office of Governmentwide Policy. Washington, D.C.: General Services Administration.

Hamel, G., and C.K. Prahalad. 1994. Competing for the Future. Boston, Mass.: Harvard Business School Press.

International Association of Machinists and Aerospace Workers. 1999. Fact Sheet Summary of Drafts of HR 716 / S 314 "Freedom from Government Competition Act." Available on line at: *http://www.iamaw.org/departments/cobarg_govaffairs.htm*

Johnson Controls. 1997. Testimony of Roy Cloudsdale, Vice President Global Business Delivery Systems, Johnson Controls World Services, Inc., before the Subcommittee on Readiness of the House Committee on National Security. Washington, D.C.: Government Printing Office.

Kelman, S. 1998. Implementing Federal Procurement Reform. Innovations in American Government Occasional Paper Series. Cambridge, Mass.: Harvard University John F. Kennedy School of Government.

Khalilzad, Z.M., and D.A. Ochmanek, eds. 1997. Strategy and Defense Planning for the 21st Century. Prepared for the United States Air Force, Project AIR FORCE. Washington, D.C.: RAND.

Kiihne, D.A. 1993. Planning Your Outsourcing Strategy. Facilities Planning News. Orinda, Calif.: Tradeline Incorporated.

Krizan, W.G. 1997. Business Roundtable takes aim at nagging project problems. Engineering News-Record 239(22): 13.

Laurent, A. 1998. On time, at cost. Government Executive 30(9): 14–23.

Light, P.C. 1999. The True Size of Government. Washington, D.C.: Brookings Institution Press.

Lopez-de-Silanes, F., A. Shleifer, and R.W. Vishny. 1997. Privatization in the United States. The RAND Journal of Economics 28(3): 447–471.

Marchese, Mark. 1993. World Trade Center Disaster. EPICCGRAM Fall 1993 (#3). New York: Emergency Preparedness for Industry and Commerce Council.

Minoli, D. 1994. Analyzing Outsourcing: Reengineering Information and Communication Systems. New York.: McGraw-Hill.

Moavenzadeh, F. 1997. Risk Management. Construction Business Review 7(4): 5.

National Research Council (NRC). 1986. Programming Practices in the Building Process: Opportunities for Improvement. Building Research Board, National Research Council. Washington, D.C.: National Academy Press.

NRC. 1989. Improving the Design Quality of Federal Buildings. Building Research Board, National Research Council. Washington, D.C.: National Academy Press.

NRC. 1990. Achieving Designs to Budget for Federal Facilities. Building Research Board, National Research Council. Washington, D.C.: National Academy Press.

NRC. 1991. Uses of Risk Analysis to Achieve Safety in Building Design and Operations. Building Research Board, National Research Council. Washington, D.C.: National Academy Press.

NRC. 1994. Responsibilities of Architects and Engineers and Their Clients in Federal Facilities Development. Board on Infrastructure and the Constructed Environment, National Research Council. Washington, D.C.: National Academy Press.

NRC. 1995. Measuring and Improving Infrastructure Performance. Board on Infrastructure and the Constructed Environment, National Research Council. Washington, D.C.: National Academy Press.

NRC. 1997. Synthesis of Highway Practice 246: Outsourcing of State Highway Facilities and Services. Transportation Research Board, National Research Council. Washington, D.C.: National Academy Press.

NRC. 1998. Assessing the Need for Independent Project Reviews in the Department of Energy. Board on Infrastructure and the Constructed Environment, National Research Council. Washington, D.C.: National Academy Press.

NRC. 1998. Stewardship of Federal Facilities: A Proactive Strategy for Managing the Nation's Public Assets. Board on Infrastructure and the Constructed Environment, National Research Council. Washington, D.C.: National Academy Press.

NRC. 1999. Improving Project Management in the Department of Energy. Board on Infrastructure and the Constructed Environment, National Research Council. Washington, D.C.: National Academy Press.

Pint, E.M., and L.H. Baldwin. 1997. Strategic Sourcing: Theory and Evidence from Economics and Business Management. Prepared for the United States Air Force, Project AIR FORCE. Santa Monica, Calif.: RAND.

Project Management Institute. 1996. A Guide to the Project Management Body of Knowledge. Silva, N.C.: PMI Communications.

RAND. 1997. Defense issues: infrastructure reform, golden goose or false hope? Available on line at: *http://www.rand.org/publications/CF/CF133/*

Robinson, M., and S. Wilson. 1994. Privatization in Massachusetts: getting results. Government Union Review 15(1): 1–55.

Stahl, N. 1998. IFM: Where Microsoft Is Going Today. FM Data Monthly. Available on line at: *http://www.fmdata.com/fmdm/issues/...here_microsoft_is_going_today.html*

Staudohar, P. D. 1981. Contracting out in federal employment. Government Union Review 2(2): 3–10.

U.S. Army Corps of Engineers. 1990. A Guide to Effective Contractor Quality Control (CQC). Washington, D.C.: U.S. Army Corps of Engineers.

Wallin, B.A. 1997. The need for a privatization process: lessons from development and implementation. Public Administration Review 57(1): 11–19.

Washington, W.N. 1999. Outsourcing automatic data processing requirements and support. Acquisition Review Quarterly 6(2): 195–206.

Worsnop, R.L. 1996. Privatizing government services. Congressional Quarterly Researcher 6(30): 697–720.

Acronyms

ANG	Air National Guard
ANGRC/CEC	Air National Guard Readiness Center / Civil Engineering Center
APS	Advanced Photon Source (Project)
ASRM	Advanced Solid Rocket Motor (Project)
CM	construction manager
DOE	U.S. Department of Energy
FAIR	Federal Activities Inventory Reform Act
FAR	Federal Acquisition Regulation
FBO	Foreign Buildings Operations (Office of)
FY	fiscal year
GAO	General Accounting Office
GOCO	government-owned contractor-operated
GPRA	Government Performance and Results Act
IBB	International Broadcasting Bureau
NASA	National Aeronautics and Space Administration
NAVFAC	Naval Facilities Engineering Command
NIST	National Institute of Standards and Technology
NRC	National Research Council

OFPP	Office of Federal Procurement Policy
OMB	Office of Management and Budget
PM	program manager
SSC	Superconducting Super Collider (Project)
USACE	U.S. Army Corps of Engineers